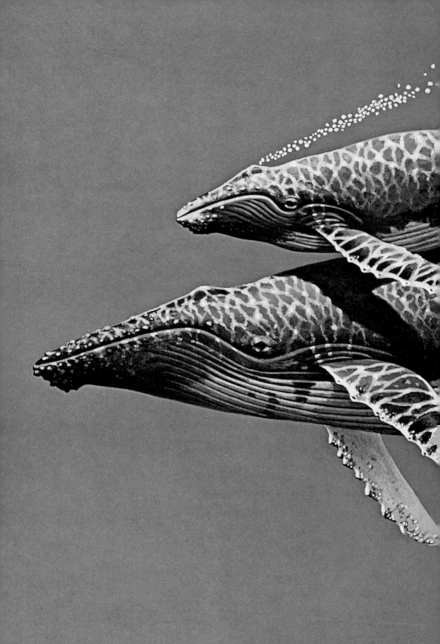

RICHARD ELLIS

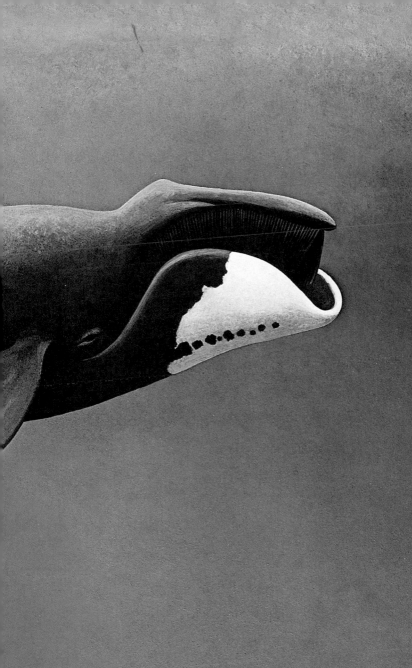

CONTENTS

WHALES
GIANTS OF THE SEAS AND OCEANS

Yves Cohat and Anne Collet

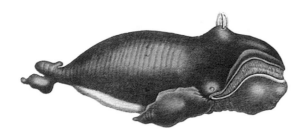

DISCOVERIES®
HARRY N. ABRAMS, INC., PUBLISHERS

Whales have always held an enormous fascination for people. Over the years an entire whale mythology grew up, inspired by the creature's vast size and the mystery surrounding its habits. It was not until the 18th century that the whale was designated as a mammal rather than a fish, while cetology as a scientific discipline and a branch of zoology dates back only to the 1960s.

CHAPTER 1
AN EXCEPTIONAL MARINE ANIMAL

Opposite: a humpback leaping out of the water off the island of New Caledonia. Scientists can still only guess at the significance of such behaviour. Right: a 19th-century Japanese illustration of a whale.

The Bible and Pliny's *Natural History*

The first reference to a whale occurs in the Bible, in the story of Jonah, although the word 'whale' does not appear. The text simply states: 'Now the Lord had prepared a great fish to swallow up Jonah.' Jonah is thrown overboard by his fellow sailors during a violent storm at sea, is swallowed by a whale and vomited three days later on dry land. Another biblical creature, Leviathan, is almost certainly a whale as well. Mentioned for making the abysses of the sea boil like a cauldron, Leviathan is referred to again in Psalm 104: 25–6: 'So is this great and wide sea, wherein are things creeping innumerable, both small and great beasts. There go the ships: there is that leviathan, whom thou hast made to play therein.'

Pliny the Elder (AD 23–79), who travelled to North Africa during the first century AD, explains in his *Natural History* that the sea is so vast that it is hardly surprising if a host of strange and monstrous creatures are found in it. He adds that in the Gallic ocean, as he terms it, a huge fish called *physeter* (the scientific name of the sperm whale, meaning 'blower' in Greek) was seen emerging out of the sea in the manner of a column or pillar, taller even than a ship's sails; and that this fish sent a huge jet of water spurting high into the air, as if the water were being ejected by a pipe.

Below: a 16th-century illustration of a 'bearded' whale.

The *Speculum Regale*

Virtually all the medieval texts on whales are Scandinavian or Icelandic. The main one, the *Speculum Regale* (mid-13th century), describes various species of whales living in the seas around Iceland. It refers to orcs having dog-like teeth and demonstrating the same kind of aggression towards fellow cetaceans as dogs do towards other land animals. According to the *Speculum Regale,* orcs would attack larger

•Then God spake unto the fish; and from the shuddering cold and blackness of the sea, the whale came breeching up towards the warm and pleasant sun, and all the delights of air and earth; and "vomited out Jonah upon the dry land"• (Herman Melville, *Moby Dick,* 1851). Left: a 16th-century lithograph showing Jonah emerging from the whale.

The word cetacean derives from the Greek *ketos*, meaning 'sea monster', and a host of monstrous and bizarre visions have grown out of the rich mythology surrounding the whale. This 16th-century illustration of a 'porcine' whale (below) suggests that the creature once walked on land.

whales in groups, biting their isolated victim until it died of blood loss and exhaustion – though not before killing a number of its attackers with a powerful blast from its blowhole.

The *Speculum Regale* describes a motley collection of creatures, including well-known species such as the sperm whale and the narwhal, together with terrible monsters, killers of men and destroyers of ships, bearing strange names like 'horse-whale', 'pig-whale' and 'red whale'.

Not all the creatures described are fierce and cruel, however. Some might even be said to be benevolent: for example, the whale that drives shoals of herrings and other fish towards the shore is particularly helpful to fishermen. It is remarkable that the creature takes care not to hurt the fishermen themselves or damage their boats, as if – the *Speculum Regale* claims – God had decreed that this state of things should continue as long as the fishermen behave in a peaceable fashion; but if fighting breaks out between them and blood is spilled, the whale blocks the way to the shore and drives all the fish back towards the open sea.

The Devil's whales and whale-islands

In the Middle Ages Icelandic sailors had a particular fear of what were termed the 'Devil's whales'. The word 'whale' was itself taboo – and any sailor who forgot the taboo found himself deprived of food – because it was feared that whales would be attracted by the mention of their name and would come and attack the ship and attempt to destroy it. The whales were instead called 'big fish'. Some, it was thought, particularly favoured human flesh and would linger

I n this 16th-century engraving (below) two men measure a beached pilot whale before the onlookers arrive.

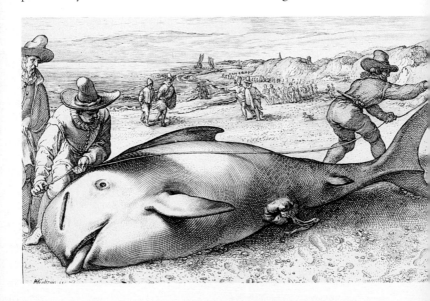

for a whole year in a spot where they had once enjoyed such a delicacy. Shallow waters where whales were known to have sunk ships were therefore places to avoid.

Stories of sailors mistaking a sleeping whale for an island are widespread in folklore. In the story the whale generally wakes up, dives and drowns the sailors. The motif reappears in the legend of St Brendan (AD 484–c. 578), the Irish Benedictine abbot who, in AD 565, sailed west across the Atlantic in search of the Holy Land. In the course of his journey he disembarked with his men on the back of a huge whale, whereupon the abbot calmly proceeded to set up an altar and celebrate Mass – without the usual tragic consequences.

Mammals or fish?

The science of 'cetology', the branch of zoology concerned with the study of whales, dates back in its most basic form to antiquity. For many hundreds of years it depended on observations of dead whales whose bodies had been washed up on a beach.

Worshippers celebrate Mass on the back of a whale (above). Such an incident occurs in St Brendan's legendary sea voyage, a symbolic account in which God is the ocean and the Church is the ship sailing on the waters of eternity. In their pursuit of Paradise the saint and his followers are subjected to various trials, including storms and encounters with a fierce sea monster. On one occasion they set foot on an island as white as a flock of sheep; on another a black and featureless island that sinks before their eyes – the back of a whale. In fact St Brendan probably reached the Faeroe Islands (the white island) and Iceland (the black island).

Over four hundred years before the birth of Christ, the Greek philosopher Aristotle (384–322 BC) identified whales as mammals. Pliny the Elder later declared them to be fish. Many naturalists continued to side with Pliny, but Pierre Belon (1517–75) and G. Rondelet (1507–66), while persisting in classifying whales as fish, describe them as having lungs and a 'matrix' or uterus, like mammals. It was not until the tenth edition (1758) of the *Systema naturae*, written by the Swedish botanist Carolus Linnaeus (1707–78), that cetaceans were finally classed as mammals. A few decades later the French zoologist and palaeontologist Baron Georges Cuvier (1769–1832) described the

CONRADI GESNERI
MEDICI TIGVRINI
HISTORIÆ ANIMALIVM
LIBER IV.
Qui eſt de Piſcium & Aquatilium animan-
tium natura.
Cum Iconibus ſingulorum ad vivum expreſſis ferè
omnibus DCCXII.
Editio ſecunda, novis formarum, natura oſervationibus non paucis melior, atque etiam multis in locis emendatior.
Conſtitutur in hoc Volumine.
GVILIELMI RONDELETII non nunquam, Medicina Profeſſoris Regii in Scholis Monſpelienſi, &
PETRI BELLONII CENOMANI, libellus ſiue compendii Latini Latini,
de Aquatilium Singulis Genere.
Paraliponeon quorum et ſuorum adderis hinc.
Magniſtim in locis Loquentibus.

[engraving of a fish/eel]

FRANCOFVRTI,
In Bibliopolio Andreæ Cambieri.
ANNO M D LXX.

Title page of the fourth book of the *Historia animalium* (1558; left) by the Swiss naturalist Conrad Gessner, one of the first to provide a description of whales. A few hundred years later, thanks to the accounts of ships' captains such as William Soresby Jr and Charles Scammon, zoologists were much better informed about the anatomy and physiology of whales, and whale skeletons like this one (below) were assembled and exhibited in museums across Europe.

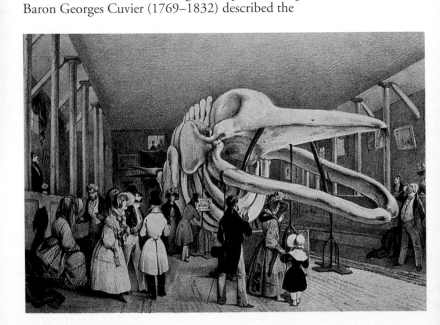

whale as a mammal without hind legs. Whale skeletons were reconstructed in the first museums of natural history, and comparisons with fossils of extinct land mammals led zoologists to the conclusion that cetaceans are an ancient family of mammals that probably share a common ancestor with land mammals.

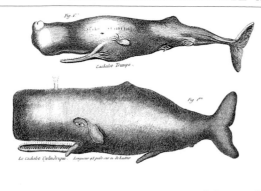

The long evolution of cetaceans

It is now known that whales were originally adapted for life on land. However, due to the scarcity of fossil evidence, it is not possible to trace the exact evolution of the whale in its passage from land-based to aquatic mammal.

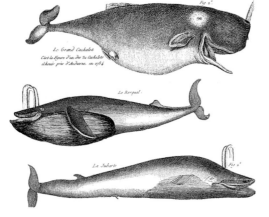

Palaeontologists recognize the mesonyx as the ancestor of the first cetaceans. A carnivorous land animal similar in appearance to a wolf, with a long muzzle and four hooved feet, the mesonyx lived during the Eocene epoch, fifty-five million years ago, in the marshy areas bordering the coast, where it fished for food.

Over the course of a few million years its limbs shortened, its muzzle lengthened and its nostrils shifted position, ending up on top of the animal's head in order to facilitate breathing when the head emerged out of water.

Illustrations such as these (above), taken from a work by the French zoologist and palaeontologist Georges Cuvier, were based on second-hand information: accounts of stranded whales or observations made at sea. Cuvier concluded that there were three species of sperm whale, whereas there is in fact only one.

The most remarkable structural change, however, involved the disappearance of the animal's pelvis and hind limbs, resulting in a body perfectly adapted to movement through water. A horizontal flipper was activated by the powerful muscles of the tail stock or caudal peduncle, and it was the upward motion of this caudal peduncle that propelled the animal through the water and enabled it to ascend to the surface to breathe.

Organs such as the ears, the male genitalia and the mammary glands – normally protruding in land mammals – became internalized as part of the process of change. The animal's skin became smoother and an increased fat reserve provided protection against the cold. Eyes and ears were adapted to facilitate underwater vision and hearing, and the nostrils formed a blowhole on top of the head.

Although whales may look like fish, they are in fact mammals. Fish adapt to the temperature of their surroundings and breathe by absorbing dissolved oxygen through their gills; whales are warm-blooded creatures equipped with lungs for filtering oxygen out of the air. Fish lay eggs; whales give birth to live

The mouths of whalebone whales have long rows of baleen plates that have a smooth outer surface and a hairy fringed inner side (above) for filtering and trapping food.

Below: the fossilized skull of an 'archeoceti', showing that the ancestors of the whale had teeth, and nostrils found towards the tip of the animal's muzzle. These creatures, which lived in shallow seas and estuaries fifty million years ago, are the oldest known marine mammals.

offspring developed in the uterus, and then suckle their young for several months.

Whalebone whales

The order of the *Cetacea* is divided into two suborders: whalebone whales (*Mysticeti*), or true whales, which have strips of whalebone in place of teeth and a double blowhole, and toothed whales (*Odontoceti*), which have numerous small teeth and a single blowhole. New genetic techniques are broadening our knowledge all the time, leading to increasingly detailed classifications: by mapping the genes of individual whales, biologists are constantly establishing new relationships between species and families. Until recently, for example, only one species of minke whale was thought to exist, but new research has shown that the Antarctic population in fact forms a species apart.

Whalebone – found only in whalebone or baleen whales (rorquals, right whales and grey whales) – is composed of keratin (which also occurs in hair and nails), a fibrous, elastic material in the form of plates

The eyes of all large cetaceans are located on the sides of the animal's head, close to the corner of its mouth. This grey whale (above) is feeding, tilting its head to the left and looking at the photographer with its right eye. Grey whales feed on marine crustaceans such as sand hoppers, which live in the sandy deposits of shallow waters. Swimming on one side (grey whales generally favour their right side), they sift the sand with their rubbery lips, then suck in the tiny organisms disturbed by this activity before filtering out the water and sediment through their whalebone.

hanging from the upper jaw on either side of the palate. There are huge numbers of these plates (as many as 800 in the blue whale), tightly packed together. They create the effect of a fringed curtain that acts as a filter for sifting out krill and small fish from the seawater taken into the whale's mouth.

Four families of whales are currently recognized: right whales or *Balaenidae*, which include four species (the bowhead whale, the North Atlantic right whale or Basque whale, the North Pacific right whale and the southern right whale); the rorquals or *Balaenopteridae*, with seven species (the humpback whale, the common minke whale, the Antarctic minke whale, Bryde's whale, the sei whale, the fin whale and the blue whale); the *Neobalaenidae*, with one species (the pygmy right whale); and finally the *Eschrichtiidae*, with one species, the grey whale, which has similarities with both the right whale and the rorqual. The blue whale is the largest of all whales, measuring up to 36 m (118 ft) in length and 160 tonnes in weight.

Above: a grey whale and her calf. Grey whales often carry parasites. Below: by contrast, this North Atlantic right whale is covered with clusters of crustaceans.

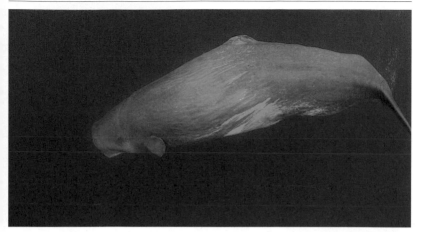

Toothed whales

The other suborder of cetaceans, the toothed whales (*Odontoceti*), comprises nearly seventy species divided into ten families. The best known are the *Delphinidae* (dolphins, pilot whales and killer whales), the *Phocoenidae* (porpoises), the *Monodontidae* (narwhal and white whale), the *Physeteridae* (sperm whale) and the *Kogiidae* (pygmy sperm whale and dwarf sperm whale). There are also several families of freshwater dolphins: the *Iniidae*, the *Lipotidae*, the *Pontoporiidae*, the *Platanistidae,* and the *Ziphiidae*, which include twenty little-known species such as the beaked whale and the bottlenose whale. Toothed whales have a single blowhole, while whalebone whales have a double blowhole.

The male sperm whale (*Physeter macrocephalus*, meaning 'large-headed blower') can be 18 m (59 ft) long, but the female is much smaller, not exceeding 12 m (39 ft). The sperm whale can be recognized by the size of its head, which comprises a third of its body mass. Its forty or so teeth are located in the long, narrow lower jaw; conical and as long as 18 cm (7 in), they fit into corresponding cavities in the upper jaw when the animal's mouth is closed.

Above: a sperm whale swimming in the waters off the Azores. Sperm whales are able to dive for exceptionally long periods – over an hour – without surfacing for a breath of air, a feat which, despite the physiological changes already recognized, biologists are still at a loss to explain. One means of saving energy is achieved by diving vertically and returning to the surface in the same manner (thus using the fastest route). Sperm whales feed on giant squid, locating their prey using their highly developed sonar, then patiently lying in wait for it on the ocean bed.

Krill – the whale's favourite food

Whales feed principally on plankton, a huge carpet of living organisms that is suspended in the upper levels of the sea and drifts back and forth with the currents. In amongst these minute organisms, a combination of plants and animals, are shrimplike marine crustaceans called krill, the whale's favourite food,

which is absorbed in vast quantities after passing through the sieve-like filter of the animal's whalebone.

Towards the end of winter whales migrate to the icy waters of the polar regions, where krill are to be found in abundance, undertaking a journey of between 3000 and 8000 km (about 1800 and 5000 miles). The few brief months of the polar summer provide the whales with their main source of sustenance for the entire year, and an individual whale will absorb several tonnes of krill during this period, in a real feeding frenzy. The massive right whales travel at no more than 5 knots (9 kph or 5½ mph), but the slighter and more powerful rorquals can achieve speeds of up to 15 to 20 knots (28 to 37 kph or 17 to 23 mph). Some whales can 'porpoise'

The term krill covers a number of different species of small crustaceans that vary depending on the locality. In Antarctica the most commonly found is *Euphausia superba* (below), which is between three and six cm long. During the feeding season a whale consumes several tonnes of krill daily, as this fin whale's stomach (above left) shows.

Phytoplankton (the plant constituent of plankton) forms the lowest level of the marine food chain and depends for its development on the presence of light and various nutrients found in the sediment on the ocean bed, where the water is coldest. For the phytoplankton to grow, these icy waters, laden with nitrates, need to float to the surface where the light can penetrate. Such conditions occur in the polar regions and in areas of so-called 'upwellings', along coastal shallows, where oceanographic factors combine to produce ascending currents of glacial water (in Peru and Chili, and off the west coast of Africa, for example). These cold upper waters provide the environment for a particularly dense and rich biomass, and whales spend the summer eating tonnes of krill from Arctic seas – such as here (above left) at Spitsbergen – dotted with icebergs. The continuous light (since the sun never sets) encourages plankton to proliferate. Feeding humpback whales (following pages) frequently attract a flock of sea birds, which come to pick off some of the krill disturbed by the whales' activity.

(make little leaps on the surface of the sea) as a way of minimizing strain over short distances (since air provides less resistance than water), and during long migrations whales reduce their speed to no more than 3 to 5 knots on average.

The only mammals to reproduce at sea

Whales leave the icy polar seas for warmer waters at lower latitudes in order to mate and reproduce. They are sexually mature somewhere between five and twelve years, but only begin reproducing between the ages of eight and fifteen years.

Gestation lasts between ten and fourteen months – depending on the species – and the mother suckles her young for a minimum of six months and then rests for a while before starting again. A whale will thus give birth every three or four years during the most fertile period of its life.

Twins are extremely rare. They tend either not to reach full term or to die soon after birth because the mother is unable to produce enough milk to sustain two offspring at a time. A whale's life expectancy varies between thirty years for the smaller species and almost a hundred for the larger ones. Infant mortality is fairly high among cetaceans and it is estimated that whales produce a maximum of eight young that will themselves reach reproductive age.

Unlike most land mammals, cetaceans are born tail first to enable the animal to swim straight to the surface for air as soon as its head disengages. After taking its first few breaths the whale calf returns to its mother to suckle. In mammals lips are normally an indispensable part of suckling, but in cetaceans a different process operates. The female's mammary glands are surrounded by a muscular ring and when the calf makes contact with the slit concealing the gland, her teat becomes erect and the gland is activated in such a way as to spurt milk directly into the calf's mouth.

The milk of marine mammals contains as much as 50 per cent fat, high levels that are necessary for the rapid weight gain of the young calf, which is born with only a thin covering of fat. It is estimated that the female blue whale can

No one has ever witnessed the birth of a whale in its natural habitat, but the process – which lasts approximately half an hour – has been filmed with captive dolphins, as in this photograph (above left). The umbilical cord ruptures when the head emerges, and the calf is immediately pushed up to the surface so that it can take its first breath of air. When the mother is hunting, another female takes her place with her young.

produce around 500 kilos (1100 pounds) of milk a day, enabling her calf to put on approximately 100 kilos (220 pounds) a day during the first weeks of life.

Breathing, diving and sleeping

'It's spouting! It's spouting!' On ships around the world, century after century, the same cry could be heard as the ship's watch spotted another whale from his eyrie at the top of the mast. In fair weather the whale was visible from several miles away.

Cetaceans are able to exhale air very rapidly due to the shape and the relatively small size of their lungs, and also to the horizontal position of the diaphragm. Right whales produce a double spout, in the form of a V, whereas rorquals expel a single column of spray of variable height and volume depending on the species.

Whale calves are lighter in colour than adults at birth, but their pigmentation rapidly darkens. This young humpback whale (above, with its mother) can scarcely be more than one week old, judging by its colour. During their early months the young calves are often 'carried' on the back of their mother – or, more accurately, on the currents created by the adult's movement through the water – and can thus keep pace with their mother, despite their small flukes.

Record diving times (with a corresponding holding of breath) are held by the sperm whale, a number of toothed whales of the *Ziphiidae* family (beaked whales) and various seals. Sperm whales have been found trapped by telephone cables at depths of more than 1000 m (3280 ft), and recent soundings have located groups of sperm whales at a depth of almost 3000 m (9840 ft). By contrast, whalebone whales rarely dive below a few hundred metres, remaining below the surface for a maximum of forty or fifty minutes. Breathing patterns of course vary according to the species and size of the animal, and according to its activity at the time. For the most part, however, breathing is suspended for no more than ten to twenty minutes, followed by an interval of several minutes during which the whale returns to the surface, where it will inhale and exhale some six to fifteen times.

When a whale sleeps, it remains virtually motionless just below the surface of the sea, coming

The grey whales of the eastern Pacific spend the winter in the Sea of Cortez in Baja California, Mexico, where they reproduce; then, in spring, they undertake the lengthy journey to the Bering Sea, off Alaska, to feed. Once virtually hunted to extinction because of commercial whaling, grey whales are now strictly protected. Aeroplanes fly over the length of their route so that their numbers can be calculated each year, and, despite the authorized capture of a handful of grey whales by the Yupik of Alaska and Siberia, the number of whales has increased to more than 25,000.

up for air every ten minutes on average by kicking gently with its caudal peduncle. For land mammals, sleep is a reflex action: they can sleep deeply without thinking about breathing any more than they have to think about making their heart beat or digesting their food. For cetaceans, by contrast, respiration remains a voluntary act. Since the whale has to 'remember' to breathe, even when it is asleep, there is no risk that it will automatically breathe underwater when the levels of carbon dioxide rise in its blood during a prolonged dive. It is thought that when a whale sleeps only one hemisphere of its brain is resting, while the other remains partially active in order to control the animal's breathing, and that, after a period of activity, this hemisphere shuts down and the other takes over.

Communication in whales

In human beings communication depends primarily on language, whereas animals rely on a number of other factors, in particular body language and

All whalebone whales have a double blowhole – two quite distinct nostrils – whereas toothed whales have a single blowhole on the top of their head. Only right whales, however, blow out a double jet of spray, as this whale is doing (above). Rorquals emit a single jet whose shape and height varies according to species and to meteorological conditions. With the exception of a handful of species with a particularly characteristic spout (right whales and sperm whales), only a specialist is capable of identifying a whale by its spout.

behaviour. We frequently miss
the significance of animal noises,
because, instead of looking at the
context in which they occur and
the behaviour accompanying them,
we pay attention to the sounds.

Whales emit sounds in a band of
frequencies varying from 20 hertz
to 3000 hertz, favouring the lower
frequencies that travel long
distances underwater. Rorquals have been recorded at
a distance of 30 km (over 18 miles) from the micro-
phone, and acoustic studies demonstrate that their
singing can in fact be heard at a distance of several
hundred kilometres, further than 1000 km (620
miles), even, in certain oceanographic conditions.
Rorquals live alone or in small groups and use fairly
short, monotonous sounds to maintain contact
between individuals, particularly at night or in the
course of lengthy migrations.

The singing associated with whales is produced by
the humpback whale, and research has focused on
this species more than any other. Only the male of
the species sings, mainly in winter when the animals
are preparing to reproduce, repeating short melodies
sometimes for hours on end. These songs differ from

In order to study a
group of animals, it is
important to be able to
distinguish individuals
within the group. The
study of whale
behaviour is based on
photo-identification,
whereby each animal
is identified according
to individual morph-
ological characteristics.
Right whales tend to be
classified by the clusters
of small parasitic
crustaceans that form
white, grey or yellow
patches on the animal's
skin, as on the head and
around the eye of this
right whale (above). In
the case of humpback
whales the clearest
distinguishing feature
is the pigmentation
of the tail flukes (left).
The colour can vary
from completely black
to completely white,
with or without
orange markings.

one whale population to another and are developed over the course of a season; but it takes years for the musical phrase to change beyond recognition from its initial starting point. Each animal appears to possess its own musical signature, enabling it to be identified at a distance. Members of the same group also share certain sequences, and a tonality and accent.

Humpback whales leap out of the sea with such force that their whole body clears the surface of the water for a few seconds before falling back in a huge mountain of spray. These movements

Whale songs appear to serve a variety of purposes. They can be a means of marking out territory; they can be a part of the courting ritual; an expression of defiance towards rival males; and a way of communicating a message (a call to a partner, a warning signal to the rest of the group, or an expression of sexual availability, for example). Central to these various functions is the need for identification: the singer communicates to a potential audience his or her sex, age, status, activities and emotional state. The significance of a melody, and the effect it is designed to produce, will depend on the context in which it is used.

clearly play a part in communicating with other whales. They are not necessarily intended to transmit a message in themselves, but perhaps serve as a kind of physical exclamation mark emphasizing a previous vocal signal. The young calves engage in a great deal of leaping, mimicking the adults for whom these movements seem to carry great significance.

NAVTÆ IN DORSA CETORVM, QVAE INSVLAS PVTANT,
*anchoras figentes fæpe periclitantur. Hos cetos Trolual sua lingua
appellant, Germanis e Teuffelwal.*

SIMILIS EST ET ILLORVM ICON APVD EVNDEM, CAPITE,
*roftro, dentibus, fiftulis, quos montium inftar grandes effe fcribit, & naues euertere, nifi fono tubarum aut miffis in
mare rotundis & vacuis vafis abfterreantur: quod & in Balthico mari circa
balenam Brunsfich dictam fieri diximus.*

Whales were hunted and slaughtered by the Basques in the early Middle Ages, and from the 9th to the 16th century the Basques monopolized the trade in whale products. Whaling was a lucrative business and attracted the Dutch and the English, but it was the colonists of New England who were the most serious hunters. The golden age of North American whaling, which began in 1835, lasted a quarter of a century.

CHAPTER 2
WHALING

There have been numerous illustrations of whales over the centuries. Opposite: a 16th-century engraving showing a whale being flensed and a ship being attacked by 'monsters'. Right: a 17th-century engraving of a whale killed at Spitsbergen.

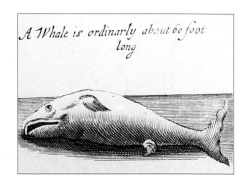

A Whale is ordinarly about 60 foot long

The Basques: the first whalers

While it is impossible to give a precise date for whales first being hunted, it is likely that whale carcasses washed ashore in prehistoric times would have been recovered for their meat.

Archaeological digs in Japan have uncovered harpoons and animal bones that indicate that small cetaceans were hunted during the Jomon period from 3000 BC. In Norway there is evidence to show that whales were diverted back from shore into the narrow gulleys of the fjords, where they were then speared to death – a technique known as the *grind*.

From the 9th century onwards whale hunting was an important part of life for the Basques. For six months of the year the Bay of Biscay would be full of whales, or *sardas*, as they were known to the Basques – North Atlantic right whales (*Eubalaena glacialis*) that had flocked to the warm waters of this sheltered corner of the Atlantic Ocean in order to give birth to their young.

Every autumn lookouts would be posted in hilltop towers overlooking the sea and signal the approach of a group of *sardas* by beating drums, ringing bells and

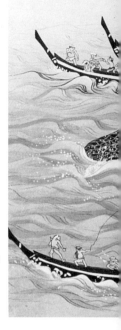

Fig. 2.

lighting fires. The hunters would then put to sea in skiffs and row out towards the whales, guided by those watching from the clifftops. The crew of ten consisted of rowers, harpooners and a helmsman. As the boats drew alongside the whales, the men attacked them with spears; the dead whales were then towed back to shore, where their carcasses were cut up.

Whaling soon became a thriving industry. Trading outlets opened up on the French and Spanish coasts, dealing in a whole range of whale products: meat, blubber, oil and also the much-prized whalebone. *Sarda* tongue was regarded as a delicacy and sold in the markets of Biarritz, Bayonne and Ciboure, and salted whale blubber could be bought in any French town.

It is thought that the Japanese began hunting whales in their coastal waters as early as 3000 BC. Above: a specialized method of hunting, known as the *amitori*, practised in Japan in the 16th century. It involved driving the whale into shallow water and trapping it in a net.

Opposite: in the 17th century the Basques were renowned for their whaling skills and initiated foreign sailors in the art.

By the 15th century, however, *sardas* began to desert their temporary home off the coast of northern Spain. Either they had grown wary of the hunters or their numbers had been seriously depleted by over-hunting. Whatever the reason for their disappearance, whales no longer flocked to the Bay of Biscay in sufficient numbers to sustain the former trade. The Basques responded by extending their hunting grounds further north, using bigger merchant ships known as carracks, which sat high in the water and measured 20 metres (65 ft) or more in length. The crews for these ships, originally from the ports of Normandy, were simply employed to convey the hunters to their destination, the Basques reserving for themselves the 'noble' tasks of harpooning and

Above: a whale being flensed alongside a ship in Spitsbergen. The flensers stood on the carcass of the whale itself, attached by safety ropes, and cut off strips of blubber that were then hoisted aboard using a large winch.

flensing. A little later the carracks were replaced by caravels, two- and three-masted ships that were faster and more seaworthy, and carried the hunters in pursuit of their quarry as far as the Faeroe Islands, off the coast of Iceland. This new type of whaling required different processing techniques, and carcasses were now cut up at sea. Inevitably, however, the whale blubber began to decompose and stink as the ships sailed south into warmer waters. The solution was to melt the fat on board ship and to store it as oil in barrels stowed in the hold.

The Dutch initiative: whaling as an industry

The 16th century was a turning point for whaling. With France and Spain engaged in endless conflicts,

If a whale was captured close to a coastal station, the carcass was towed in and flensed on shore. The strips of blubber were then cut into pieces and liquefied in circular brick melting pots. The resulting oil was cooled in special vats half full of water before being decanted into barrels. Above: scenes showing the melting and decanting processes.

the Basques could no longer sustain the commercial success they had hitherto enjoyed. Meanwhile, two other countries, Holland and England, were struggling for control of the major sea lanes and equipping their own fleets of whaling ships.

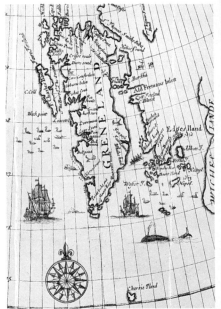

At the outset both the English and the Dutch employed harpooners and flensers from the Basque country – the only nation that had so far made a 'profession' of whaling. Having learnt the necessary techniques, however, they dismissed their Basque employees and even barred them from sailing in northern waters. A desperate battle ensued, with both the Dutch and the English attempting to gain the monopoly over hunting for bowhead whales.

In the early years of the 17th century the Dutch began building the first whaling stations along the coast of Spitsbergen (Svalbard), a Norwegian archipelago in the Arctic Ocean. Apart from the melting houses and shops and depots connected with the industry itself, a whole variety of other shops and small businesses grew up in these small towns. Each had its inns, and even a church, and teemed with life during the whaling season, reverting to a ghost town once the whalers had left.

Whales flocked to these waters off the coast of Spitsbergen, and whaling here was a precisely organized operation. The men would go out, six to a boat, and when they were sufficiently close to a whale, they would plant a harpoon in its side. The harpoon was attached to the boat by a long rope, with the result that when the whale swam off, in great pain, it dragged the boat along behind it. This was the most dangerous part of the operation: the men had to be prepared to give the whale more rope if it suddenly dived; otherwise they were forced to sit tight and allow themselves to be pulled along. When the animal began to show signs of exhaustion, the boat would approach and the whalers would finish it off with spears and drag the carcass back to shore.

The carcass remained where it was for a day or two before the flensers began their work, removing the blubber and heating it to extract the oil. The barrels of oil were then transported by raft to the mother ship for delivery to the Dutch ports.

It was not long, however, before the right whale populations were exhausted by such systematic killing. The whales had come to feed on the dense carpets of krill in the waters off Greenland; now they migrated much further west, towards Labrador, off the northeastern coast of Canada. Their departure signalled the demise of the great Dutch whaling stations.

For many years the coasts of Spitsbergen were confused with those of Greenland, as this early 17th-century map (opposite) shows. In 1623 the small town of Smeerenberg (below, in a painting from 1634) sprang up almost overnight on Spitsbergen's inhospitable coastline. Smeerenberg, also known as 'blubber town', was built exclusively to meet the needs of the whaling industry and out of season it resembled a ghost town. At its height, in the 1630s, it employed as many as 300 whaling ships and housed between 12,000 and 18,000 seasonal inhabitants.

The Nantucket whalers in pursuit of the sperm whale

Around 1650 the English colonies of North America began developing their own small-scale whaling industry, using lookouts posted along the coast to signal the arrival of whales, as the Basques had done before them.

By 1670 the New England industry was flourishing and their products were being exported to both England and its other colonies, and to the rest of Europe. Ports specializing in whaling were scattered all along the coast, from Long Island to Cape Cod. After 1690, however, a single port – Nantucket, an island off southeast Massachusetts – became the centre of the New England whaling industry.

In 1712 Christopher Hussey, a ship's captain from Nantucket, was hunting for right whales when he found that he had sailed into waters inhabited by a species traditionally feared by whalers: the sperm whale. This fear was based on the fact that sperm whales are predators (hunting for food rather than grazing), equipped with teeth rather than whalebone,

To hunt whales, the Kodiak Indians (above), inhabitants of the Aleutian Islands (south west of Alaska), poisoned the tips of their harpoons with aconite and, having wounded the whale, left the poison, the winds and the currents to do the rest, simply waiting for the carcass to be washed ashore. The tribe that had discovered the washed-up carcass would examine the wound in an effort to identify the tribe responsible for the slaughter (since the spearhead would bear a characteristic mark). The relevant village was then alerted and the booty shared between the two tribes.

and that they swim in huge herds, sometimes several hundred together, organized in a quasi-military manner. If a whaler caught sight of a herd of sperm whales in the distance, he would turn his ship around rather than confront the animals. On this particular day in 1712, however, Christopher Hussey did not turn his ship around; instead he attacked one of the whales and brought the first dead specimen back to land. It was a feat of daring that had repercussions for the whole Nantucket whaling industry and marked the beginning of hunting on the open sea.

In 1715 Nantucket had six whaling sloops (single-masted vessels). In 1775, at the beginning of the War of American Independence, it was a flourishing whaling station whose wealth was guaranteed by its success in hunting sperm whales. The oil of the sperm whale is superior to that of true whales, and the sperm whale also provides spermaceti, a substance extracted from its head and used in the manufacture of candles, and ambergris, which is secreted by its intestinal tract and used in the manufacture of perfumes.

After 1715 American whalers concentrated their efforts on hunting sperm whales because of their high market value, a single animal providing 10,000 litres of oil. In normal conditions the sperm whale swims at a speed of 3 knots, but the American whalers found that whales attempting to evade capture could maintain a speed of 10–12 knots for an hour at a time. Migrating sperm whales undertake a lengthy circuit in the company of their mates and young (above), and once the hunters were familiar with these routes, success was guaranteed, leading to the slaughter of 20,000 whales between 1842 and 1846.

The annual production of the New England colonies at this time was 45,000 barrels of spermwhale oil, 7500 pounds of whalebone and 8500 barrels of baleen (whalebone) whale oil.

The War of American Independence was a disaster, however, for New England's whaling industry. The English navy attacked its ports and either captured or sank its ships, and sailors who were taken prisoner were offered the choice of serving on English whaling ships or being bundled into the hold. The 'American', Dutch and French fleets were out of the running, leaving the field wide open for their British rivals.

The supremacy of the English and Scottish whaling industry

In 1756 the English and Scottish whaling fleets together numbered 83 vessels. Twenty years later, thanks to the war, they had extended their field of activity to include the whole of the Atlantic, from the Arctic seas around Spitsbergen down as far as the east coast of South America via the coast of Canada.

In 1790 the *Emilia* cleared Cape Horn and finally returned from a highly fruitful expedition in the Pacific with its holds literally stuffed with spermaceti and whale oil. The English were swift to consolidate this success. Over the course of the next three years they sent a total of twenty-three ships to the Pacific, and, on the strength of their success there, established whaling ports in Australia and New Zealand.

In the early years of the 19th century, however, the English whaling industry suffered several losses that led to the end of its

When the War of American Independence broke out in 1775 Timothy Folger, a wealthy Nantucket merchant (below), was one of those who resisted the general tide of feeling against Britain. In 1785 he led a group of 300 whalers loyal to the British Crown to Nova Scotia and started up a fishery there. In 1792 the British persuaded the 300 loyalists to move to the port of Milford Haven, in Wales, in an effort to help build up the British whaling fleet and counter the growing success of the port of Nantucket.

monopoly. Whales were increasingly scarce in the Atlantic, and English whaling ships provided ideal targets for Napoleon's frigates. Meanwhile, in the Pacific, an American vessel, the *Essex*, sank twelve English ships. After 1815, with a peace agreement in place between England and America (following the War of 1812), the reorganization of North American ports led to renewed interest in the Pacific route, and in the 1820s more than 120 American vessels sailed back and forth across the waters of the Pacific.

Once it had recovered from the after-effects of the war, America devoted huge efforts to the revival of its whaling industry. Many American sailors who had been forced to enlist on English ships returned to New England, and in six years, from 1835 to 1841, the North American whaling fleet – sailing out of thirty ports – increased the number of its vessels from 203 to 421. In 1846 New England and New York State numbered some fifty or so ports and 735 ships. Each port had what might be regarded as its own speciality: the sailors of New London hunted right whales; those of Stonington, right whales and Arctic whales (or bowhead whales); and the small vessels of Provincetown, Atlantic whales. The two ports that would be most readily remembered in the history of whaling, however, were Nantucket and New Bedford.

The *Charles W. Wanderer* (above, in a painting of 1923 by C. W. Ashley), the pride of the New Bedford fleet and the last whaling ship powered entirely by sail, was driven on to rocks 20 km (13 miles) from port and smashed to pieces on 26 August 1924. During its eighty-year service the *Charles W. Wanderer* had undertaken 37 expeditions and brought its owners profits of almost 2 million dollars.

Nantucket's golden age

Around 1820 Nantucket was North America's major whaling port, its wharves crumbling under the weight of its barrels of oil – and the price of this oil continued to rise, matched by the ever-increasing demand for whalebone.

Factories for processing oils and whalebone sprang up throughout the town, alongside a thriving industry devoted to the manufacture of sails and harpoons, metal cords and rivets. Around 1830 seventy-two whaling ships left Nantucket each year, bringing back 30,000 barrels of precious oil that was exported worldwide and used to lubricate the machines of the new Industrial Revolution and illuminate the cities of Europe and America.

The growth of Nantucket's whaling industry was limited, however, by geographical considerations, as a sandbank at the mouth of the harbour restricted the size of the boats that could enter. Meanwhile, larger and larger whaling ships were being used to exploit the new hunting grounds in the Pacific, and Nantucket gradually lost its supremacy as a whaling port as New Bedford supplanted it.

Above: New Bedford in southeast Massachusetts, a leading whaling port during the 19th century.

Opposite: a corner of the sail loft, a painting by C. W. Ashley from 1915.

New Bedford

From 1842 onwards America was unrivalled in terms of its whaling industry. Of the 882 whaling ships employed around the world, 652 were American. In one year alone, 1851, New Bedford equipped 249 ships. Seven years later its ships returned to port with oil and whalebone amounting to a total value of six million dollars.

The whaling industry had brought great wealth to the little town of New Bedford and the average income of its inhabitants – natives of Africa, the Azores and the South Sea Islands, virtually all of whom depended on the industry in one way or another – was one of the highest in the world. There was visible evidence of this prosperity, from the quaysides humming with activity to the stately residences gracing the hills. The reek of the candle-making factories permeated the town,

Above: a 19th-century engraving showing barrels of oil being checked and counted at the port of New Bedford. A successful expedition produced some 1500–2000 barrels, a single sperm whale providing between 20 and 40.

• The sail loft on Front Street in New Bedford is cool and airy, even on a hot day.… The sail-makers sit quietly in their aprons, stitching the heavy cloth with triangular sail needles and palms. (A *palm* is a piece of leather fitted to the palm for protection and equipped with a metal slug for pushing the needle through the cloth.) In a suit of sails for *Ulysses* there are hundreds of thousands of stitches, eleven thousand square feet of canvas, and miles of thread with easily forty pounds of beeswax on it.•

Jan Adkins
Wooden Ship, 1978

In 1841 an American artist, Benjamin Russell, set sail from New Bedford on the ship *Kutussoff*. He served as a cooper on this whaling vessel for three years and five months, during which time he recorded the various stages of the hunt in the most intimate detail. On his return he decided to recapture the entire voyage on canvas and collaborated with a house painter, Caleb Purrington, to produce a *Panorama of a Whaling Voyage Round the World* (1849). The panorama, measuring almost 500 m (over 1600 ft), was unrolled by hand on stage while a narrator simultaneously described the action. This scene (left), showing the American fleet hunting right whales in the north Pacific, is one of Russell's most famous. To the right of the picture a sperm whale has just been harpooned, while to the left of the picture another sperm whale has attacked and capsized a whaleboat. When the holds were full of oil and whalebone, the ships returned to New Bedford to sell their cargo.

Following pages: hunting whales and bears off the coast of Greenland.

competing with the musky scent of ambergris
destined for the perfumeries of London and Paris.
In the huge rope factories teams of workers were busy
assembling the cords for harpoon guns, and at every
hour of the day or night the taverns were full of
sailors, some of them recently returned from a
whaling expedition, others on the point of departure.

The decline of traditional whaling

The decline of the North American whaling industry
began in the second half of the 19th century. Whale
numbers were decreasing and whaling expeditions
were extended further and further afield, frequently
yielding nothing.

The discovery of gold in California on 24 January
1848 provoked a literal gold rush. Increasing
numbers of sailors deserted ship, turning from
whalers to gold-diggers overnight.

The discovery of American crude oil, a decade
later, aggravated the situation. Crude oil swiftly
replaced whale oil for lighting purposes and within
a few years the price of whale oil had plummeted.

During the American Civil War (1861–5) whaling
ships were either sunk or captured by corsairs from
the Confederate States; ships' companies were
disbanded and sailors forced to enlist on southern

From June onwards
Californian grey
whales are to be found
feeding in the icy waters
of the Bering Strait,
where they are capable
of breaking up ice more
than a metre deep – an
achievement that earned
them the nickname 'ice-
breakers' among former
American whalers.

ships. By the time the war was over whales had deserted the coasts of America and whalers – those of them who remained – were obliged to venture further and further north in search of their quarry. Unlike their British rivals, recently converted to steam power, American whalers continued to rely on sail, which left them vulnerable to the elements, particularly to the fierce conditions prevailing in the Arctic. In 1871 thirty-four ships in the American flotilla were trapped in the ice floes off the coast of Alaska. The incident was

In August 1871 forty American whaling vessels were stationed in the Bering Strait when the weather suddenly deteriorated and a section of the fleet found itself trapped in the ice. It was a financial disaster for the American whaling industry, with losses estimated at some 3 million dollars. Left and below: paintings by Benjamin Russell illustrating the disaster.

a tragedy for the port of New Bedford. Gradually new stations were created at the instigation of local shipowners on the west coast of America (in San Francisco Bay and Monterey Bay) and came to replace the moribund ports of New England.

In the last years of the 19th century American financiers themselves lost interest in the whaling industry, preferring to invest their capital in a new and expanding area, cotton; and whaling as a large-scale industry slowly began to die out.

In the mid 19th century a whaler spent an average of forty-two months at sea. '*There* is his home', as Herman Melville said: 'For years he knows not the land.' He alternated between joy and tedium, elation and loneliness, swapping the delights of a tropical island for a violent encounter with a foreign people. To survive such a life required enormous courage and powers of endurance.

CHAPTER 3

OF WHALES AND MEN

There were few distractions on a whaling vessel and the expeditions were long and at times tedious. Engraving whale bones and teeth, like this one (right) from a sperm whale, was a favourite pastime. Opposite: *The Gam*, 1926.

The Pacific route

From 1850 to 1860 the majority of America's whaling fleet was mobilized in the Pacific. Sperm whales were still the whalers' most prized quarry, and there were probably more than a million of them swimming in Pacific waters at that time.

The whaling route varied according to the time of year, the known migrations of the whales and the captain's whim. Generally speaking, however, the ships passed Cape Horn to reach the Pacific in October or November and then sailed on to the Galapagos Islands, continuing from there to the Sandwich Islands (the former name of Hawaii) and from there to Japan.

They scoured the waters of the Pacific until their holds were bursting with barrels of oil. They would put into port in one of the islands, in order to stock up with provisions, and might stay there for several months, even a whole year, before returning home via Australia, the coast of Africa and the Atlantic – thereby circumnavigating the globe.

The captain and his men

On board ship the captain exercised total authority, answering to no one except the ship's owner on his return to port. East of Cape Horn he owed his allegiance to God and to his employer; west of Cape Horn he was, as he liked to think, God in his own right – and he often ruled the ship with a heavy hand.

The captain needed excellent navigating skills and a familiarity with the habits of whales and their whereabouts. His expertise covered the various stages of the hunt: the search, the approach, the chase, the harpooning, the engagement with the whale and the flensing of the carcass. He might take charge of one of the whaleboats or, alternatively, stay on board the mother ship in order to direct

operations from there. As there was no doctor on board, the captain dressed wounds and set fractures.

He was supported by the first mates, who supervised all the work done on board and were responsible for the smooth running of the ship. It was also their job to finish off the harpooned whale with spears and to oversee the flensing and melting down of the blubber.

The harpooners were an essential part of the crew, because the success of the campaign depended on their skill. The cooper made the barrels for storing the oil and made sure that they were properly stowed in the hold. He would assume command of the ship if the captain and his first mates were in the boats or in the event of an emergency. The rest of the crew consisted of the steward, who served the captain and his officers, the carpenter, the cook and the sailors who made up – in addition to the officers and harpooners – the four-man teams to man the boats.

The Pacific route was first navigated in 1789 and in the 19th century the majority of vessels using it were American. The map (above) shows the route taken by the *Kutussoff*, which sailed out of New Bedford in 1841 and returned three and a half years later after circumnavigating the world. Also marked are the areas in which right whales and sperm whales are concentrated.

Opposite: Captain Howland, who led several successful whaling expeditions between 1827 and 1844.

The Pacific Ocean, with its warm winds and crystal-clear waters, must have seemed like paradise to the American whalers, even though August and September could bring hurricanes with wind speeds of up to 300 kph (185 mph). They must have particularly enjoyed stopping at one of the islands and supplementing their meagre ship's fare with fresh fruit – making up for the monotony and deficiencies of a diet that consisted, typically, of dried meat, rice, biscuits, beans and sometimes whale brain rolled in flour and fried in whale oil. These stopovers were rare, the temptations to desert considerable. The ship's captain would make a point of warning his crew against native customs and the danger of cannibalism, but, despite the risk, the Pacific islands (above left and left, from Caleb Purrington and Benjamin Russell, *Panorama of a Whaling Voyage Round the World*, 1849) attracted a fair number of deserters eager to exchange the harsh conditions of life on board ship and the brutality of their superiors for the delights of a palm-fringed beach.

Ships' companies recruited from all over the world

In the early years of whaling, except when there was a shortage of manpower and a small number of Indians or black people were recruited to make up a ship's complement, the crew of a whaling ship was composed of white men from New England. For these men the hunt was simply a temporary means of earning a living, though the more skilful or experienced occasionally took up whaling as a way of meeting the needs of the family back home.

After 1820 the requirements for manpower had grown to such an extent that men were recruited from all over the world and from all walks of life: Basques and Portuguese, Polynesians, Melanesians and natives of the Sandwich Islands – tough and spirited individuals, appreciated for their harpooning skills. They were joined by a motley bunch of adventurers from France, Germany, Ireland, Italy, Sweden and Spain. Many of the recruits, unaccustomed to the harsh conditions of whaling life, deserted at the first port of call. Yet a large number remained – excellent sailors and hunters, who won a reputation for courage and their unflagging efforts. For such men whaling was a passion, an irrepressible urge that led them to abandon their homes and their families for years.

The long wait

The initial weeks at sea were frequently monotonous: there were many miles to travel before the whalers reached their hunting grounds. The only distraction lay in training the new recruits in the art of harpooning and manoeuvring the boats. This training

Articles of engraved whale bone and whale tooth (often decorated with whaling scenes) are known as scrimshaws. Below: an ivory cribbage board.

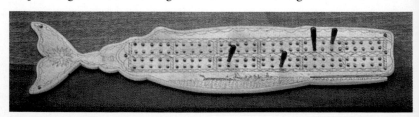

To counteract the tedium of the voyage the members of a ship's crew would often keep a journal while waiting for the hunt to begin. Left: two pages from the ship's journal on the *William Baker* containing an account of nine days during an expedition in November 1838. The illustration in the top right-hand corner shows a whaleboat broken in two by a blow from a whale's tail.

Below: a harpooner from Nantucket, c. 1820.

was rigorous and unremitting, because it was crucial that each person on board knew his job.

The wait could last weeks, sometimes months. Among the sailors a favourite way of passing the time was to produce carvings out of whale bones and the teeth of sperm whales, using a simple pocket knife. They generally engraved designs on these objects with the kind of needle used for sewing sails and then darkened the designs by rubbing them with a mixture of oil and lampblack. These creations ranged from a simple pin to luxury objects of extraordinary complexity and delicacy, and demonstrated that some of their creators were not only whalers but also artists of considerable skill.

Far left: a square-rigged three-master, like the *Pequod* in *Moby Dick*. A whaling ship was a large, heavy vessel with a hold big enough to house several hundred barrels of oil. The ship was slow, but sturdily built. It was either a square-rigged three-master or, more often, a three-master barque, having the foremasts rigged square and the aftermast rigged fore-and-aft – a lighter form of rigging that could be handled by six men, freeing up the rest of the crew to man the whaleboats.

Left: a whaleboat and harpoons. A whaleboat was light (450 kg or 990 pounds), fast and highly man-oeuvrable. It was long and narrow – approximately 9 m (30 ft) by 2 m (6½ ft) – and, being pointed at both ends, could be rowed in either direction. It was designed to carry six men, including the helmsman who steered the boat using the stern-oar. The harpoons (below) were kept in the bows, attached to several metres of cord coiled in buckets.

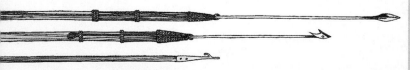

The whale's tragic end

Once the ship had reached its destination, the pace of life and routines changed dramatically. The sails were stowed at night and only set again at daybreak. Every two hours throughout the day a new lookout took up his position on a small platform near the top of the mast. When he spotted a whale, he would shout 'Spouting ... It's spouting!' and he was able to identify which species of whale it was from its spout. A whale's spout generally reaches a height of 3–4 m (10–13 ft) and lasts for approximately three seconds. It can be seen within a radius of six miles. At this distance it is easy to make a mistake and, in order to avoid unnecessary commotion on the deck, a lookout might wait until he was absolutely certain of a sighting before giving the signal. If the whale projected its head out of the water he would shout 'Jumping ... It's ju-u-umping!', and 'Sounding ... It's sou-ou-nding' to indicate that he had seen the whale dive.

As soon as a whale had been glimpsed, the boats were launched. The captain gave the order to brace

This painting by C. W. Ashley (1909; above) shows what was known as the 'Nantucket sleighride', the period of the hunt when a whaleboat and its occupants were dragged at great speed (at least for the first few hours) by the harpooned whale. The crew had to try and keep as close to the wounded animal as possible and give out more line if it dived. The cord, which was coiled in large buckets in the middle of the boat, passed from stem to stern around a curved stanchion, the 'drum', before returning to the bows on a roller protected by a cleat.

the sails of the mainmast and bring the ship to.
Then the chase began, either under sail or using
oars, depending on the winds and sea and the
distance to be covered. The officer acting as
helmsman, or headsman as he was known, steered
the whaleboat using an oar almost 7 m (23 ft) long.
The harpooner rowed with the other oarsmen in the
bows until they were close enough to the whale for
him to act. Then, with his two arms raised, he held
the harpoon in both hands and directed a powerful
but carefully aimed blow at the whale's back. Once
the harpoon was firmly embedded in the animal's
flesh, the helmsman shouted 'All speed astern' and
the oarsmen rowed the boat away from the wounded
whale as swiftly as possible. The creature's reactions
could be violent and it was not unusual for boats
to sink at this juncture because the men had not
succeeded in distancing themselves quickly enough.

If the harpooned whale continued to swim on the
surface, the whalers allowed it to drag the boat behind
it while they attempted to check its speed as far as
possible. If, however, the whale decided to dive, the

A helmsman finishes
off a wounded
sperm whale after
exchanging places with
the harpooner (above).
In the background the
crew of the mother ship
are flensing a whale
carcass lashed to the
side of the vessel.

One of the great
dangers of whaling
was the risk of being
struck by the tail of a
wounded animal; or of
the boat capsizing when
the whale resurfaced
after a dive, as illustrated
(overleaf) in this famous
watercolour by Louis
Garneray (1835).

men let out more line, doubling the length – if necessary – by using cord from another boat. If the line ran out before they had a chance to act, they simply attached pieces of wood to the end and left everything to drift in the hope of recovering it later.

As the whale began to tire, the men pulled themselves closer by hauling on the cord. The helmsman then changed places with the harpooner in preparation for the 'noble' task of dispatching the whale. Searching for the part of the body known to whalers as the 'life', he struck the animal a number of powerful blows with a spear. Once he was sure that he had dealt the decisive blow, he ordered the oarsmen to put a few metres' distance between them and the creature, which continued to lash the water dangerously before finally giving up the struggle and rolling on to its side in a lake of blood. It usually took about ten blows to dispatch the whale. Death was caused by a pulmonary haemorrhage, and the men knew that it was imminent when the animal's spout was reddened with blood – an emission known as 'red roses'. A pole carrying the whaleboat's flag was planted on the carcass – to enable the body to be located later – and the hunt continued.

Flensing, a highly complex operation

Once the hunt was over, it was time to harvest the spoils. The body   of each whale in turn was lashed

The dead whale was hauled to the starboard side of the ship and literally 'skinned' (opposite). Pieces of whale blubber were piled on the deck, where a ship's boy would slice them into thin sheets, rather like the pages of a book, in order to accelerate the process of liquefaction when the 'bibles', as they were known, were melted down. The resulting oil was poured into special vats to cool, then transferred to barrels and stored in the hold. By the end of the operation the stench of oil was such that some sailors claimed that you could smell a whaling ship for miles around. The process was identical for a sperm whale. Once they had removed all the blubber, however, the flensers used their long knives to hunt in the animal's entrails for that precious commodity peculiar to the sperm whale – ambergris.

lengthways to the side of the ship, with its head
directed towards the stern and its tail secured by
a chain. A special platform with a parapet was
then positioned directly above the whale, and it
was here that the flensers set to work cutting up
the animal using long-handled flensing knives with
curved blades, blubber spades and blubber cutters.

During the flensing of a sperm whale the lower jaw
was removed and the spinal column severed, then the
huge head was separated from the rest of the body
and transferred to the ship's stern while the crew
continued to cut up the rest of the animal and melt
down the blubber. When these processes were
complete, the head was positioned at deck
level. A hole was made in it and the spermaceti
was carefully collected into a bucket and
immediately boiled to prevent it from
decomposing. It was then stored in barrels,
which were lined up alongside the bulwarks for
three days to allow the oil to cool. The whale's
massive carcass was abandoned to the sharks.

Opposite: this sketch
of 1855 shows
a group of men using
a hoist to remove the
teeth of a sperm whale.

The invention of the harpoon gun in the 19th century led to the destruction of several populations of large whales. It was almost a century before the International Whaling Commission intervened effectively, and today, despite a total ban on commercial whaling, several species are still under threat.

CHAPTER 4

FROM PERSECUTION TO PROTECTION

While Japan continues to exercise the right to hunt whales (opposite, in Antarctica), the practice is generally regarded as inhumane and has provoked widespread protest around the world, including these schoolchildren (right) in Glasgow in June 1992.

The invention of the harpoon gun

In 1864 a technical innovation revolutionized whale hunting – the invention of a formidable new harpoon by the Norwegian Svend Foyn, a successful seal hunter and ship's captain. The harpoon, which was fired by a gun and could travel a distance of 50 m (164 ft), exploded on contact. Once it became embedded in an animal's flesh, the barbs opened out into a star shape, breaking a vial of sulphuric acid, and this in turn caused the powder chamber to catch light. Death ensued for the whale in a matter of minutes, whereas the original harpooning technique could lead to hours of suffering.

Foyn had a new ship, the *Spes et Fides*, specially constructed, with his harpoon gun installed in the bows, and over the course of a season's hunting, in 1868, he killed thirty rorquals – a feat unprecedented in whaling history. Rorquals are fast swimmers and had not been hunted hitherto. They are also

Svend Foyn (left), inventor of the harpoon gun – a harpoon and launching device in one – which relegated the old hand-held harpoon and spear of traditional whaling to the museum. This gun (below), made in Norway in 1925, is equipped with a sight and fires a large harpoon with a pointed head – a design since superseded by one with a conical head.

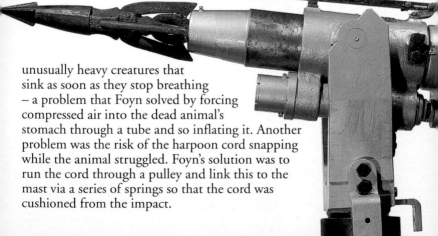

unusually heavy creatures that sink as soon as they stop breathing – a problem that Foyn solved by forcing compressed air into the dead animal's stomach through a tube and so inflating it. Another problem was the risk of the harpoon cord snapping while the animal struggled. Foyn's solution was to run the cord through a pulley and link this to the mast via a series of springs so that the cord was cushioned from the impact.

Norway, Russia and Japan: the whaling nations

The Norwegians knew better than anyone where the highest concentrations of rorquals were to be found in Arctic waters. Five Norwegian companies began operations in Iceland, and in 1889 the largest whaling station ever built was constructed at Onundard-fjördhur on the northeast coast. Over eleven years 1296 whale carcasses were processed there. In 1891 Russia and Japan began to develop flourishing industries of their own. A peculiarity of Japanese stations was that they had no slipways – unlike the Norwegian stations – for processing the carcasses, which were flensed from the ship's side.

Meanwhile, whale populations were diminishing off the Norwegian coast and Norwegian whaling companies were encouraged by the government to extend their activities outside territorial waters. Initially this action led to the creation of new stations no further north

Up until the 1960s the Norwegians and the Japanese hunted out of the port of Grytviken on the island of South Georgia (a dependency of the Falklands in the South Atlantic). More than 800 people were employed there – processing the blubber, canning and drying meat, preparing bonemeal and repairing the boats. Today Grytviken is a ghost town, slowly rusting into oblivion, but the air there is still tainted with the bitter tang of whale oil. Above: the rusting remains of a whale catcher, a small vessel equipped with a harpoon gun which brought carcasses back to shore for processing.

than Iceland, the Faeroe Islands and the Hebrides. Gradually, however, Norwegian activities spread to the coast of Labrador, and Newfoundland, where eighteen whaling stations were already in existence in 1905, Spain, Morocco, the Pacific coast of North America, Africa and South America.

1914–45: sonar tracking

The Norwegians first constructed factory ships, enormous vessels that sailed from one site to another and, casting anchor in a sheltered bay, carried on the work of a whaling station where it was impossible to do so on land. The Norwegians also improved Foyn's harpoon gun. By replacing the vial of acid with a delayed-action mechanism, they increased the depth of penetration and also made it easier to load the gun.

The beginning of the First World War did nothing to check the expansion of Norway's whaling industry; rather the contrary, thanks to the increasing demand for glycerine, extracted from whale oil and used as a component in explosives. A number of new instruments were invented, including an electric

A whaling station employed between 200 and 300 men, including the manager and foreman, naval engineers, blacksmiths, warehousemen, mechanics, cooks and men responsible for flensing the whales. A team of men could dismantle a whale in an hour and a half, and an average of twelve whales were processed a day. Above: Norwegian flensers holding the long-handled knives used for stripping a carcass.

harpoon, a flat-headed harpoon and a device for launching the harpoon like a rocket. It was the Japanese, however, who produced the most sophisticated invention – the use of a sonar to indicate the direction in which whales were swimming and their distance from the boats.

From the Arctic to the Antarctic, whales of every size and species, both adults and young, males and females, were tracked in this way. Twelve thousand large cetaceans were slaughtered in 1920; 15,000 in 1922; 17,000 in 1924; 27,000 in 1926; and 44,000 in 1931, including more than 30,000 blue whales harpooned in Antarctica.

In 1934 the Japanese – with their sights now set on Antarctica – bought a Norwegian factory ship, the *Antarctic*, and five hunting vessels. Renamed the *Tonan Maru*, this factory ship and its flotilla returned to Japan, hunting en route. From 1937 to 1938 the Japanese constructed four whaling stations in Antarctica. Two years later they began whaling on the open sea in the North Pacific.

The spinal column of a large whale (below) lying on a beach in Patagonia. Such dismal relics are not uncommon along the shores of earlier hunting grounds. In cold regions, where humidity is reduced by the action of the winds, biodegradation is an extremely slow process, and massive bones like those pictured here take several hundred years to disintegrate totally.

The Second World War gave whales a brief respite from hunting, but in 1945 nine factory ships were again scouring the waters of Antarctica – six of them equipped by Norway, two by Great Britain, and one by South Africa. In 1946 the number of ships had increased to eleven, seven of them Norwegian, three British and one South African. Meanwhile, Japan was authorized by the Allies to rebuild its whaling fleet.

The International Whaling Commission

The respite occasioned by the war had not been sufficient to repopulate the seas, and whale numbers

continued to diminish. In 1946 an international commission met in Washington, D. C., to consider a solution. Nineteen countries, including a handful in favour of whaling (Japan, South Africa and the Soviet Union), ratified an agreement that aimed to control the hunting of whales and set up the International Whaling Commission (IWC).

At its first meeting in London, three years later, the IWC established a system of quotas using the blue whale as a unit of measure. According to this system 1 blue whale was equivalent to 2 fin whales, 2½ humpback whales and 6 sei whales. The right whale and the grey whale were both declared 'protected

A blue whale (above left) photographed in the Atlantic in August 1985. The blue whale is the largest living creature on earth. The females (which are larger than males in whalebone whales) reach up to 35 m (115 ft) in length and weigh up to 150 tonnes. In the water, where it is not always easy to appreciate its length, the whale can be identified by its rounded muzzle.

species'. Moreover, the IWC banned the capture of whale calves and females in calf of any species at all.

However, this system of evaluation was innately flawed, because it was based on quantitative rather than qualitative notions and failed to take account of the comparative rarity of each species. If, for example, a country had a quota of 10 units, it could indiscriminately slaughter 10 blue whales, or 8 blue whales and 4 fin whales, or 60 sei whales. In 1971 the system was abandoned and a new strategy adopted. This time the quotas were calculated by species, but these calculations were still based on a limited knowledge of the state of individual populations.

Whales gather in large numbers off the coast of Iceland in summer. Until the beginning of the 1980s they were vigorously hunted in these waters and brought back to the coastal stations for processing. Whales have three separate stomachs (mechanical, chemical and pyloric). Above: the mechanical stomach of a fin whale at Hvalfjördhur in Iceland.

The Soviet Union and Japan continue the slaughter

During the 1960s whaling diminished because of the cost of the expeditions and a shortage of manpower. Great Britain, the Netherlands and Norway stopped sending ships to Antarctica. The first two completely disbanded their whaling fleets; the third kept only a few vessels for operations along its own coastline. With the discovery of substitutes for a number of whale products, whaling no longer had a significant role to play in a modern economy. The oil from the jojoba plant, for example, had come to replace spermaceti oil, and a chemical fixative was now used in place of ambergris.

The only countries where whaling remained a profitable activity were the Soviet Union and Japan, and for precise reasons. In the Soviet Union whaling supplied the shortfall in the chemicals industry, as the country was incapable of producing sufficient quantities of high-quality synthetic oils. In Japan whale meat was an integral part of the national diet.

During the 1961 hunting season a total of 1200 southern right whales were slaughtered by the Soviet Union, despite the fact that right whales were a protected species. Two years later the Soviet Union slaughtered more than 500 blue whales, though it only declared a fraction (75) of this figure.

A moratorium on commercial hunting

A call for a ten-year moratorium on whaling received support from a large majority at a United Nations conference in 1972, but the whaling nations chose to ignore the proposal. The plight of the whales began to excite public opinion, and there was widespread

A solidified mass of ambergris (above), formed by the concretion of undigested material in the intestinal tract of sperm whales. When first extracted, ambergris is soft, dark and sickly smelling, but with exposure to light and air its texture hardens, its colour fades and it loses its unpleasant odour. Ambergris has been found to act as an excellent fixative in the manufacture of perfumes. Sold by weight, it cost more than gold in the 18th century.

Opposite: a Japanese whaling vessel in action in Antarctica. Whale meat is regarded as a delicacy in Japan and is served in smart restaurants (where whale steak can cost even more than the best caviar) and sold by retailers, sometimes in the form of tinned meat, as here (left).

support for the concerns of the IWC. Non-governmental organizations such as Greenpeace began campaigning for an end to commercial whaling. At Washington, D. C., on 3 March 1973, twenty-one states signed the most important international agreement on the protection of living species set forth by the Convention on International Trade in Endangered Species (CITES). In the same year Japan and the Soviet Union violated their quota agreements and slaughtered over six thousand whales.

It was in 1982, following a proposal made by France, that the IWC finally adopted a moratorium on commercial whaling. The moratorium came into force in 1986 and banned all forms of commercial exploitation. The member countries of the IWC were entitled to object to any decision made by the IWC if they deemed that such a decision ran counter to their national interests. Japan did object to the decision, but then, following a bilateral agreement with the United States in 1984, withdrew its opposition, proposing instead to adapt its commercial fleet

Greenpeace was one of the first organizations to campaign for a total ban on commercial whaling. Protesters would engage directly with ships on the open sea, endeavouring to interpose themselves between the vessel and the hunted whale. In this photograph (below), taken off the coast of Spain at the beginning of the 1970s, the ship has just fired and the harpoon has reached its target: the Greenpeace protesters have been obliged to retreat to safety and watch helplessly as the whale thrashes out its last agonizing moments.

for the purposes of scientific research. Whaling for scientific purposes – in other words, as a means of gathering scientific data on different species – was authorized by the IWC. Only Japan was granted a licence for such scientific research, and its activities were limited to Antarctica, and to the capture of one hundred minke whales annually. According to the IWC's ruling, however, there was nothing to prevent Japan selling the remaining whale meat, as frozen stocks, once the necessary samples had been removed.

A recent study conducted by scientists from America and New Zealand reveals that more than a third of Japan's minke whales had been taken from the Sea of Japan, an area where this creature is scarce.

In 1993 the Norwegian government unilaterally declared that it was resuming commercial whaling in its territorial waters in the North Atlantic, thereby

Above: a dying whale. The bloody emission from its blowhole indicates that its lungs have been injured. The blood of whales contains much higher concentrations of haemoglobin than the blood of land mammals, and a wounded whale will make a huge expanse of water completely red.

Watched by a group of schoolchildren proudly waving flags, a Japanese factory ship (left and above) leaves its home port to hunt minke whales in Antarctica in November 1998. The Japanese continue to regard whaling as a noble activity, a tradition dating back to a time when their ancestors pitted themselves against giants using the smallest of weapons. Whales are no match, however, for contemporary factory ships, which are equipped with sonars to detect the position of diving whales, harpoon guns, winches and a rear ramp for hauling the animal up on to the bridge.

Overleaf: Japanese whalers flensing a minke whale on a factory ship in Antarctica in 1993.

maintaining its objection to the moratorium adopted
by the international community seven years earlier.

The IWC finally recognized the need to designate
an area of the globe where whales could enjoy total
protection, and in 1994 a sanctuary was created in
the South Pacific. This zone, situated in latitude
40° South, was henceforth subject to much stricter
surveillance than had been observed after the
moratorium vote in 1986.

On 13 April 2000, at the eleventh session of
CITES in Nairobi, Japan reaffirmed its opposition to
the measures for protecting endangered species and
continued to reserve the right to commercial whaling
of its grey whales and minke whales, whose annual
sales are officially valued at 140 and 500 units.

Immunity for indigenous peoples

The moratorium relates exclusively to commercial
whaling. Each year the IWC allocates quotas to so-
called indigenous populations, such as the Inuit of
the Canadian Arctic and Greenland, the Yupik of
Siberia and South Alaska, and certain Caribbean

Indians. These peoples have been hunting whales for thousands of years and still eat whale meat today – thus achieving a measure of self-reliance and a lack of dependence on government subsidies. Such circumscribed hunting as theirs is not considered instrumental in the destruction of whale populations.

The quotas set from 1998 to 2002 for people living in the Arctic regions are 280 bowhead whales and 140 grey whales. The Inuit of western Greenland are entitled to 19 fin whales a year over the same period, and a maximum of 175 minke whales over five years, whereas the Inuit of eastern Greenland are limited to 12 minke whales per year. For the inhabitants of St Vincent and the Grenadines in the Caribbean the maximum quota is 2 humpback whales per year for the period from 2000 to 2002. However, these quotas have been criticized as being too high.

The Arctic races continue to employ traditional methods to hunt whales and still use hand-held harpoons, as this Yupik from Alaska (opposite) is doing to hunt a bowhead whale in the Bering Sea. They will sometimes resort to the use of motor boats rather than kayaks, however, to pursue a whale and bring the carcass back to shore. Left: this group of Canadian Inuit have just cut up a beluga or white whale (a species of toothed whale). All the edible parts of the whale are divided between the members of the group according to a strict system based on ancestry. The blubber, or *muktuk*, is regarded as a great delicacy, and a few chunks of it are consumed on the spot, either still warm or already frozen.

Overleaf: whaling as a traditional pursuit is still practised today in Indonesia. A young sperm whale is dragged by villagers in June 1999 on to the beach at Lamalera, where the entire village will help to cut up the carcass.

Whale-watching: an economic alternative?

A number of whaling countries have given up the practice of commercial whaling and adopted instead a new form of tourism – whale-watching.

Whale-watching started in California in 1955, focusing exclusively on grey whales. The activity gradually spread to other regions of the United States, then, in the 1970s, to Canada and Mexico, and since the 1980s there has been a tremendous growth in this type of ecological tourism. In 1994 there were more than five million whale-watchers all over the world. In Iceland whale-watching has tripled during the last three years, attracting more than 27,000 tourists in the 1998 season alone. In 1997 tourism in Iceland generated ten times the annual revenue that whaling had brought the country in the early 1980s.

The practice of whale-watching has caught on in other countries as well – Argentina, southern Australia, South Africa (around the Cape), Portugal (the Azores) and New Zealand, and, more recently,

Several species of whale come to reproduce in the waters off the East Coast of America – a habit that almost led to their extinction when whaling in New England was at its height. Now, thanks to a new form of tourism – whale-watching – the whales have become a valuable economic asset and their protection is assured. Regulations have had to be introduced to avoid repeatedly disturbing the same animal and to ensure that the whales are not over-exposed to close scrutiny. Above and opposite: watching humpback whales in New England.

France and a number of other Mediterranean countries.

On the Caribbean islands of Dominica and St Lucia whale-watching is already a thriving industry, responsible for the development of a whole new infrastructure (hotels, restaurants and shops) and the creation of new opportunities for local employment.

In the light of this extraordinary development some system of control is clearly needed. It is essential that a number of basic rules are respected so that the whales are not disturbed. There must, for example, be a limit to the number of boats that are allowed to sail in any given area, and also a limit to the speed at which they can travel, and the boats must be required to maintain a minimum distance from the whales themselves.

A world without whales?

According to the IWC the populations of certain species have been reduced to 5 to 10 per cent of their original numbers.

Since 1967, when a ban on hunting blue whales was imposed, the numbers of these creatures have not increased; instead they have fallen from 230,000 in 1930 to some

10,000 today. The population of humpback whales was thought to total around 115,000 prior to their exploitation; today there are only 15,000, despite the total ban on hunting humpbacks since 1966.

The dramatic reduction in these populations is due to a number of factors: pollution, maritime traffic, the reduction in food sources and a poor rate of reproduction. By contrast, the population of grey whales in the eastern Pacific, after verging on extinction, is today estimated at between 20,000 and 26,000, that is, 90 per cent of its original numbers – a phenomenon that scientists are at a loss to explain.

Right whales have been hunted without respite from the 8th to the 20th century, and the fact that they have survived at all is a miracle. We do not know how many right whales there were when they were at their peak, but at present Arctic and Antarctic populations total only a few thousand.

The essential balance of the world's oceans

It is no longer possible to calculate the total number of large cetaceans on a worldwide basis, but these creatures undoubtedly represent a considerable percentage of the biomass in the world's oceans.

Whales, like all large predators, play a crucial role in the ecological balance of their environment. This balance is fragile and complex and all its elements are interdependent, linked one to the other like the parts of a great machine. If one part is removed or shut down, there is a risk that the whole operation will be brought to a standstill.

At the dawn of the third millennium, we know very little about whales, despite their growing popularity: scientists are still unable to explain many aspects of their biological make-up and much of their behaviour. Their ocean home itself remains largely a mystery world. In view of such ignorance, how are we to provide whales with effective protection and preserve the ecological balance upon which they depend?

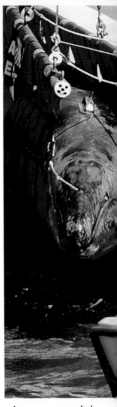

A young grey whale, Jiji', is released in the Pacific in the spring of 1998 (above). She had been found stranded on Marina del Rey beach, in California, on 11 January 1997 and had spent a year in captivity at San Diego's Sea World.

U. S. COAST GUARD

Many nations recognize the need to safeguard the world's delicate ecological structures, but, in order to understand the fundamental relations that exist between living creatures and their environments, we need to do more research, and few nations will invest sufficient funds. And yet, if we can find the weapons to destroy certain environments, we ought to be able to find the tools necessary to rebuild them. The first step is to become aware of the damage that has been done to our environment. By protecting the natural world we should realize that we are also protecting our own interests. Let us hope that our sense of self-preservation will prompt us to act, and to act soon.

Jiji's carers made up a special milk for her and fed her up to 400 kg (880 pounds) of fish per day, thanks to which her size and weight increased from 4 m (13 ft) and 875 kg (1925 pounds) on arrival to almost 8 m (26 ft) and several tonnes at the time of her release. During her most active growth period Jiji was growing by a kilogram and more than a centimetre per hour.

DOCUMENTS

The whale in literature

Whaling as a way of life has captured the imagination of some of the greatest writers of the 19th and 20th centuries. Such battles of men and giants provide the very stuff of poetry and drama.

The *St Enoch*

The next day, 7 November 1863, the *St Enoch* was towed out by the *Hercules* at high tide and quitted Le Havre. The weather was not very good. A strong southwesterly was blowing and sending a low cover of ragged clouds racing across the sky.

Captain Bourcart's vessel weighed in the region of five hundred and fifty tons and was furnished with all the usual equipment necessitated by the demands of whaling in the distant waters of the Pacific. Already some ten years old, it was nevertheless a seaworthy vessel built to withstand unpredictable weather conditions. The crew had always prided themselves on their careful maintenance of the ship, ensuring that both the hull and the sails were in perfect order, and the vessel had recently been cleaned.

The *St Enoch*, a square-rigged three-master, carried a square foresail, a mainsail and spanker, a main topsail and a fore-topsail, a main topgallant and a fore-topgallant and a mizzen topsail, a main-royal and a fore-royal, a mizzen topgallant, a fore-topmast staysail, a main jib and an inner jib and a flying jib, studding-sails and staysails.

While he was waiting for the signal to depart, M. Bourcart had installed the haulage equipment. Four canoes were in position: on the port side those of the first mate, second mate and third officer; on the starboard side the captain's. Four spare boats were stationed on the deck spars.

The stove for melting the blubber had been installed in front of the main hatch, between the foremast and the mainmast. It consisted of two adjacent iron pots built into a brick surround. The pots were heated from beneath via two furnaces situated at the front, just below the lip of each, and the smoke was vented through two holes let into the back of the pots.

The list of officers and crew entered on the *St Enoch*'s books reads as follows:

Bourcart (Evariste-Simon), captain, aged fifty;

Heurtaux (Jean-François), first mate, aged forty;

Coquebert (Yves), second mate, aged thirty-two;

Allotte (Romain), third officer, aged twenty-seven;

Ollive (Mathurin), boatswain, aged forty-five;

Thiébaut (Louis), harpooner, aged thirty-seven;

Kardek (Pierre), harpooner, aged thirty-two;

Durut (Jean), harpooner, aged thirty-two;

Ducrest (Alain), harpooner, aged thirty-one;

M. Filhiol, doctor, aged twenty-seven;

Cabidoulin (Jean-Marie), cooper, aged fifty-two;

Thomas (Gilles), blacksmith, aged
   forty-five;
Ferut (Marcel), carpenter, aged thirty-
   six;
eight ratings;
eleven apprentices;
one chief steward;
one cook.

Thirty-four men in total, the regular
complement of a whaling vessel of the
St Enoch's tonnage.

The crew divided roughly into two
halves, one half Norman, the other
Breton. The carpenter, Ferut, was the
only Parisian on board. Originally from

the district of Belleville, he had worked
as a stage hand in various of the capital's
theatres.

The ship's officers had already sailed
with the *St Enoch* on a previous
expedition. They were thoroughly
worthy men, possessing in abundance
all the qualities demanded by their
profession. During that last expedition
they had scoured the northern and
southern waters of the Pacific. It had
been a felicitous voyage, since, in the
entire forty-four months of its duration,
not a single serious incident had
occurred. A fruitful voyage too, since
the ship had brought back two thousand
barrels of oil, which had fetched a good
price.

Jules Verne, *Les Histoires de
Jean-Marie Cabidoulin*, 1902

The mother and her calf

So we were heading towards the oyster
bed in a 'disarmed' canoe – one relieved,
in other words, of its harpoons, spears
and cord. The five oarsmen, the captain
and I made seven people in total. We
had scarcely reached the middle of the
fairway and were just passing Cap
Cachalot when an enormous whale,
accompanied by her calf, suddenly
surfaced off our bows and showered us
with salt spray.

Oh, what a face Captain Jay made at
the sight of that whale – swimming right
under his nose, and he had no means of
catching it! No harpoons, no cord, and
no way of alerting our boats, which had
long since departed.

And yet he was incapable of simply
allowing such a splendid quarry to
escape.

'Captain, here's a spear,' shouted the
harpooner. 'I was going to shoot the pigs
with it in Togolabo Bay.'

The captain leapt to the front of the
boat and, brandishing his spear, shouted:
'Stand by, children! Stand by!'

The harpooner grabbed the stern oar
and, as he gave the order, the oarsmen
rowed and backed water, rowed and
backed water again.

Quite content to witness such a
spectacle, I folded my arms (having
no oar to pull); but first I took the
precaution of attaching my gun with
a bit of spun yarn to the bench where
I was sitting.

If the boat capsized, my gun would
be safe.

The mother whale was seemingly
unperturbed by our presence: she
cavorted about in the water, flipping
her body over and over, and using her
blowhole to lift the young calf, which
was struggling to keep up with her.

M. Jay held the spear down by his side, waiting for the right moment to strike. That moment came, and the spear pierced not the whale, but the calf.

At first I thought that the captain had misjudged his aim … but I soon appreciated his intelligence and skill. He knew that if the first blow failed to kill the mother, she would swim off and be lost to us; by killing the calf, on the other hand, we were disabling the mother, immobilizing her in a sense: rather than abandon her calf she would submit to being butchered on the spot.

And that was what happened. M. Jay was able to strike one blow, two blows, three blows, ten blows.… The monster thrashed around, spouted blood, and died … making no more attempt to pull away than if it had been pinioned by the most enormous harpoon. Thus, the admirable strength of maternal love that surpasses the instinct for survival!

So I could say at last that I had seen and touched a living whale, and done so in the heat of the battle.

I had seen it, and so close up that I was drenched in its blood. I had touched it, and so convincingly that my arm was almost crushed between it and the gunwale of the boat when, skimming the surface of the water in an attempt to get close to the injured calf, it hugged the side of the canoe, knocked down our lifted oars and, just as a sheep leaves some of its fleece on the bush it passes, left, on the grey paint of the planking, a coating of scaly black flakes shed by its skin.

The sleeve of my overcoat was plastered with these flakes. I shook them off proudly.

We abandoned, of course, all thoughts of wood pigeons and oyster beds. We planted a pennant on the back of the dead whale and returned to the ship to prepare the haulage equipment, while one of our number climbed to the top of the Oli-Maroa cliff and, using the flag placed there for the purpose, gave the agreed signal to our canoes summoning them to rejoin the *Asia*.

It took several hours to tow the whale back to the ship, whereupon we speedily set about hauling in the carcass.

A crowd of Maoris came to lend our men a hand, and the job was done before nightfall.

Hardly had the last piece of blubber been hoisted on deck than the natives' boats sped off towards the floating carcass and hauled it up on the beach. The spectacle that followed was at once comical and sickening as that swarm of naked men, armed with knives, some of them suspended from the animal's sides, others submerged in its gaping flank, slashed haphazardly at its flesh, helping themselves to enormous steaks of meat, which the women laid out on the sunlit grass.

That evening in every home, rich and poor alike, a fire was burning, and on every fire the delicacies of the day were cooking.

The feast began with cries of joy and songs improvised in honour of the whalers, and the next day the careful housewives hung pieces of meat on the poles of their *koamara* in preparation for harder times.

Alexandre Dumas
Les Baleiniers, Voyage aux terres antipodiques, 1858

Canto III

The boat which on the first assault did
　　go,
Struck with a harping-iron the younger
　　foe;
Who, when he felt his side so rudely
　　gored,
Loud as the sea that nourished him he
　　roared.
As a broad bream, to please some
　　curious taste,
While yet alive, in boiling water cast,
Vexed with unwonted heat, bounds,
　　flings about
The scorching brass, and hurls the
　　liquor out;
So with the barbed javelin stung, he
　　raves,
And scourges with his tail the suffering
　　waves.
Like Spenser's Talus with his iron flail,
He threatens ruin with his ponderous
　　tail;
Dissolving at one stroke the battered
　　boat,
And down the men fall drenched in the
　　moat;
With every fierce encounter they are
　　forced
To quit their boats, and fare like men
　　unhorsed....
This sees the cub, and does himself
　　oppose
Betwixt his cumbered mother and her
　　foes;
With desperate courage he receives her
　　wounds,
And men and boats his active tail
　　confounds.
Their forces joined, the seas with billows
　　fill,
And make a tempest, though the winds
　　be still.
Now would the men with half their
　　hoped prey

Be well content, and wish this cub away;
Their wish they have: he (to direct his
　　dam
Unto the gap through which they
　　thither came)
Before her swims, and quits the hostile
　　lake,
A prisoner there, but for his mother's
　　sake.
She, by the rocks compelled to stay
　　behind,
Is by the vastness of her bulk confined.
They shout for joy! and now on her
　　alone
Their fury falls, and all their darts are
　　thrown.
Their lances spent, one bolder than the
　　rest,
With his broad sword provoked the
　　sluggish beast;
Her oily side devours both blade and
　　haft,
And there his steel the bold Bermudian
　　left.
Courage the rest from his example take,
And now they change the colour of the
　　lake;
Blood flows in rivers from her wounded
　　side,
As if they would prevent the tardy tide,
And raise the flood to that propitious
　　height,
As might convey her from this fatal
　　strait.
She swims in blood, and blood does
　　spouting throw
To heaven, that heaven men's cruelties
　　might know.
Their fixed javelins in her side she wears,
And on her back a grove of spikes
　　appears;
You would have thought ... she had
　　another island been....

Edmund Waller (1606–87) from
The Battle of the Summer Islands

Moby Dick

For its power and the scale of its vision, nothing has ever been written on the subject of whales to compare with Herman Melville's novel Moby Dick. *Melville (1819–91) was a whaler and brought to his masterpiece not only the qualities of a first-class writer and storyteller but also a keen observation of his one great passion in life.*

Captain Ahab lost a leg in an earlier encounter with Moby Dick and has sworn to kill the hated white whale. After years of searching he has tracked him down, and for two long days the crew of the Pequod *have been battling with the whale. The men are exhausted and eager to give up the fight, but Ahab is like a man possessed. What none of them know is that this third day will be the last....*

The morning of the third day dawned fair and fresh, and once more the solitary night-man at the fore-masthead was relieved by crowds of the daylight lookouts, who dotted every mast and almost every spar.

'D'ye see him?' cried Ahab; but the whale was not yet in sight.

'In his infallible wake, though; but follow that wake, that's all.... Aloft there! What d'ye see?'

'Nothing, sir.'

'Nothing! and noon at hand! The doubloon goes a-begging! See the sun! Aye, aye, it must be so. I've oversailed him. How, got the start? Aye, he's chasing *me* now; not I, *him* – that's bad; I might have known it, too. Fool! the lines – the harpoons he's towing. Aye, aye, I have run him by last night. About! about! Come down, all of ye, but the regular lookouts! Man the braces!'

Steering as she had done, the wind had been somewhat on the *Pequod*'s quarter, so that now being pointed in the reverse direction, the braced ship sailed hard upon the breeze as she rechurned the cream in her own white wake.

'Against the wind he now steers for the open jaw,' murmured Starbuck to himself, as he coiled the new-hauled main-brace upon the rail. 'God keep us, but already my bones feel damp within me, and from the inside wet my flesh. I misdoubt me that I disobey my God in obeying him!'

'Stand by to sway me up!' cried Ahab, advancing to the hempen basket. 'We should meet him soon.'

'Aye, aye, sir,' and straightway Starbuck did Ahab's bidding, and once more Ahab swung on high.

A whole hour now passed; gold-beaten out to ages. Time itself now held long breaths with keen suspense. But at last, some three points off the weather-bow, Ahab descried the spout again, and instantly from the three mastheads three shrieks went up as if the tongues of fire had voiced it....

He gave the word; and still gazing round him, was steadily lowered through the cloven blue air to the deck.

In due time the boats were lowered;

but as standing in his shallop's stern, Ahab just hovered upon the point of the descent, he waved to the mate, – who held one of the tackle-ropes on deck – and bade him pause.

'Starbuck!'

'Sir?'

'For the third time my soul's ship starts upon this voyage, Starbuck.'

'Aye, sir, thou wilt have it so.'

'Some ships sail from their ports, and ever afterwards are missing, Starbuck!'

'Truth, sir: saddest truth.'

'Some men die at ebb tide; some at low water; some at the full of the flood; and I feel now like a billow that's all one crested comb, Starbuck. I am old; – shake hands with me, man.'

Their hands met; their eyes fastened; Starbuck's tears the glue.

'Oh, my captain, my captain! – noble heart – go not – go not! – see, it's a brave man that weeps; how great the agony of the persuasion then!'

'Lower away!' – cried Ahab, tossing the mate's arm from him. 'Stand by the crew!'

In an instant the boat was pulling round close under the stern.

'The sharks! the sharks!' cried a voice from the low cabin-window there; 'O master, my master, come back!'

But Ahab heard nothing; for his own voice was high-lifted then; and the boat leaped on.

Yet the voice spake true; for scarce had he pushed from the ship, when numbers of sharks, seemingly rising from out the dark waters beneath the hull, maliciously snapped at the blades of the oars, every time they dipped in the water; and in this way accompanied the boat with their bites. It is a thing not uncommonly happening to the whale-boats in those swarming seas; the sharks at times apparently following them in

the same prescient way that vultures hover over the banners of marching regiments in the east....

The boats had not gone very far, when by a signal from the mastheads – a downward pointed arm, Ahab knew that the whale had sounded; but intending to be near him at the next rising, he held on his way a little sideways from the vessel; the becharmed crew maintaining the profoundest silence, as the head-beat waves hammered and hammered against the opposing bow.

'Drive, drive in your nails, oh ye waves! to their uttermost heads drive them in! ye but strike a thing without a lid; and no coffin and no hearse can be mine: – and hemp only can kill me! Ha! ha!'

Suddenly the waters around them slowly swelled in broad circles; then quickly upheaved, as if sideways sliding from a submerged berg of ice, swiftly rising to the surface. A low rumbling sound was heard; a subterraneous hum; and then all held their breaths; as bedraggled with trailing ropes, and harpoons, and lances, a vast form shot lengthwise, but obliquely from the sea. Shrouded in a thin drooping veil of mist, it hovered for a moment in the rainbowed air; and then fell swamping back into the deep. Crushed thirty feet upwards, the waters flashed for an instant like heaps of fountains, then brokenly sank in a shower of flakes, leaving the circling surface creamed like new milk round the marble trunk of the whale.

'Give way!' cried Ahab to the oarsmen and the boats darted forward to the attack; but maddened by yesterday's fresh irons that corroded in him, Moby Dick seemed combinedly possessed by all the angels that fell from heaven. The wide tiers of welded tendons overspreading his broad white forehead, beneath the transparent skin, looked knitted together; as head on, he came churning his tail among the boats; and once more flailed them apart; spilling out the irons and lances from the two mates' boats, and dashing in one side of the upper part of their bows, but leaving Ahab's almost without a scar.

While Daggoo and Queequeg were stopping the strained planks; and as the whale swimming out from them, turned, and showed one entire flank as he shot by them again; at that moment a quick cry went up. Lashed round and round to the fish's back; pinioned in the turns upon turns in which, during the past night, the whale had reeled the involutions of the lines around him, the half torn body of the Parsee was seen; his sable raiment frayed to shreds; his distended eyes turned full upon old Ahab.

The harpoon dropped from his hand....

But he looked too nigh the boat; for as if bent upon escaping with the corpse he bore, and as if the particular place of the last encounter had been but a stage in his leeward voyage, Moby Dick was now again steadily swimming forward; and had almost passed the ship, – which thus far had been sailing in the contrary direction to him, though for the present her headway had been stopped. He seemed swimming with his utmost velocity, and now only intent upon pursuing his own straight path in the sea.

'Oh! Ahab,' cried Starbuck, 'not too late is it, even now, the third day, to desist. See! Moby Dick seeks thee not. It is thou, thou, that madly seekest him!'...

Whether fagged by the three days' running chase, and the resistance to his swimming in the knotted hamper he bore; or whether it was some latent deceitfulness and malice in him: whichever was true, the White Whale's way now began to abate, as it seemed, from the boat so rapidly nearing him once more; though indeed the whale's last start had not been so long a one as before. And still as Ahab glided over the waves the unpitying sharks accompanied him; and so pertinaciously stuck to the boat; and so continually bit at the plying oars, that the blades became jagged and crunched, and left small splinters in the sea, at almost every dip.

'Heed them not! those teeth but give new rowlocks to your oars. Pull on! 'tis the better rest, the shark's jaw than the yielding water.'

'But at every bite, sir, the thin blades grow smaller and smaller!'

'They will last long enough! pull on! – But who can tell' – he muttered – 'whether these sharks swim to feast on a whale or on Ahab? – But pull on! Aye, all alive, now – we near him. The helm! take the helm; let me pass,' – and so saying, two of the oarsmen helped him forward to the bows of the still flying boat.

At length as the craft was cast to one side, and ran ranging along with the White Whale's flank, he seemed strangely oblivious of its advance – as the whale sometimes will – and Ahab was fairly within the smoky mountain mist, which, thrown off from the whale's spout, curled round his great, Monadnock hump. He was even thus close to him; when, with body arched back, and both arms lengthwise high-lifted to the poise, he darted his fierce iron, and his far fiercer curse into the hated whale. As both steel and curse sank to the socket, as if sucked into a morass, Moby Dick sideways writhed; spasmodically rolled his nigh flank

Captain Ahab (Gregory Peck) killing Moby Dick in the 1956 film by John Huston.

against the bow, and, without staving a hole in it, so suddenly canted the boat over, that had it not been for the elevated part of the gunwale to which he then clung, Ahab would once more have been tossed into the sea. As it was, three of the oarsmen – who foreknew not the precise instant of the dart, and were therefore unprepared for its effects – these were flung out; but so fell, that, in an instant two of them clutched the gunwale again, and rising to its level on a combing wave, hurled themselves bodily inboard again; the third man helplessly drooping astern, but still afloat and swimming.

Almost simultaneously, with a mighty volition of ungraduated, instantaneous swiftness, the White Whale darted through the weltering sea. But when Ahab cried out to the steersman to take new turns with the line, and hold it so; and commanded the crew to turn round on their seats, and tow the boat up to the mark; the moment the treacherous line felt that double strain and tug, it snapped in the empty air!

'What breaks in me? Some sinew cracks! – 'tis whole again; oars! oars! Burst in upon him!'

Hearing the tremendous rush of the sea-crashing boat, the whale wheeled round to present his blank forehead at bay; but in that evolution, catching sight of the nearing black hull of the ship; seemingly seeing in it the source of all his persecutions; bethinking it – it may be – a larger and nobler foe; of a sudden, he bore down upon its advancing prow, smiting his jaws amid fiery showers of foam.

Ahab staggered; his hand smote his forehead. 'I grow blind; hands! stretch out before me that I may yet grope my way. Is't nigh?'

'The whale! The ship!' cried the cringing oarsmen.

'Oars! oars! Slope downwards to thy depths, O sea, that ere it be for ever too late, Ahab may slide this last, last time upon his mark! I see: the ship! the ship! Dash on, my men! Will ye not save my ship?'

But as the oarsmen violently forced their boat through the sledge-hammering seas, the before whale-smitten bow-ends of two planks burst through, and in an instant almost, the temporarily disabled boat lay nearly level with the waves; its half-wading, splashing crew, trying hard to stop the gap and bale out the pouring water.

Meantime, for that one beholding instant, Tashtego's masthead hammer remained suspended in his hand; and the red flag, half-wrapping him as with a plaid, then streamed itself straight out from him, as his own forward-flowing heart; while Starbuck and Stubb, standing upon the bowsprit beneath, caught sight of the down-coming monster just as soon as he.

'The whale, the whale! Up helm, up helm! Oh, all ye sweet powers of air, now hug me close! Let not Starbuck die, if die he must, in a woman's fainting fit. Up helm, I say – ye fools, the jaw! the jaw! Is this the end of all my bursting prayers? all my life-long fidelities? Oh, Ahab, Ahab, lo, thy work. Steady! helmsman, steady. Nay, nay! Up helm again! He turns to meet us! Oh, his unappeasable brow drives on towards one, whose duty tells him he cannot depart. My God, stand by me now!'...

From the ship's bows, nearly all the seamen now hung inactive; hammers, bits of plank, lances, and harpoons, mechanically retained in their hands, just as they had darted from their various employments; all their enchanted eyes intent upon the whale, which from side to side strangely vibrating his predestinating head, sent a broad band of overspreading semicircular foam before him as he

rushed. Retribution, swift vengeance, eternal malice were in his whole aspect, and spite of all that mortal man could do, the solid white buttress of his forehead smote the ship's starboard bow, till men and timbers reeled. Some fell flat upon their faces. Like dislodged trucks, the heads of the harpooneers aloft shook on their bull-like necks. Through the breach, they heard the waters pour, as mountain torrents down a flume....

Diving beneath the settling ship, the Whale ran quivering along its keel; but turning under water, swiftly shot to the surface again, far off the other bow, but within a few yards of Ahab's boat, where, for a time, he lay quiescent.

'I turn my body from the sun. What ho, Tashtego! let me hear thy hammer. Oh! ye three unsurrendered spires of

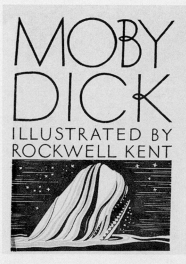

Cover of Herman Melville's *Moby Dick* in the edition illustrated by Rockwell Kent.

mine; thou uncracked keel; and only god-bullied hull; thou firm deck, and haughty helm, and Pole pointed prow, death-glorious ship! must ye then perish, and without me? Am I cut off from the last fond pride of meanest shipwrecked captains? Oh, lonely death on lonely life! Oh, now I feel my topmost greatness lies in my topmost grief. Ho, ho! from all your furthest bounds, pour ye now in, ye bold billows of my whole foregone life, and top this one piled comber of my death! Towards thee I roll, thou all-destroying but unconquering whale; to the last I grapple with thee; from hell's heart I stab at thee; for hate's sake I spit my last breath at thee. Sink all coffins and all hearses to one common pool! and since neither can be mine let me then tow to pieces, while still chasing thee, though tied to thee, thou damned whale! *Thus*, I give up the spear!'

The harpoon was darted; the stricken whale flew forward; with igniting velocity the line ran through the groove; ran foul. Ahab stopped to clear it; he did clear it; but the flying turn caught him round the neck, and voicelessly as Turkish mutes bowstring their victim, he was shot out of the boat, ere the crew knew he was gone. Next instant, the heavy eyesplice in the rope's final end flew out of the stark-empty tub, knocked down an oarsman, and smiting the sea, disappeared in its depths.

For an instant, the tranced boat's crew stood still; then turned. 'The ship? Great God, where is the ship?' Soon they through dim, bewildering mediums saw her sidelong fading phantom, as in the gaseous Fata Morgana; only the uppermost masts out of water; while fixed by infatuation, or fidelity, or fate, to their once lofty perches, the pagan harpooneers still maintained their sinking lookouts on

the sea. And now, concentric circles seized the lone boat itself, and all its crew, and each floating oar, and every lance-pole, and spinning, animate and inanimate, all round and round in one vortex, carried the smallest chip of the *Pequod* out of sight.

But as the last whelmings intermixingly poured themselves over the sunken head of the Indian at the mainmast, leaving a few inches of the erect spar yet visible, together with long streaming yards of the flag, which calmly undulated, with ironical coincidings, over the destroying billows they almost touched; – at that instant, a red arm and a hammer hovered backwardly uplifted in the open air, in the act of nailing the flag faster and yet faster to the subsiding spar. A sky-hawk that tauntingly had followed the main-truck downwards from its natural home among the stars, pecking at the flag, and incommoding Tashtego there; this bird now chanced to intercept its broad fluttering wing between the hammer and the wood; and simultaneously feeling that ethereal thrill, the submerged savage beneath, in his death-gasp, kept his hammer frozen there; and so the bird of heaven, with unearthly shrieks, and his imperial beak thrust upwards, and his whole captive form folded in the flag of Ahab, went down with his ship, which, like Satan, would not sink to hell till she had dragged a living part of heaven along with her, and helmeted herself with it.

Now small fowls flew screaming over the yet yawning gulf; a sullen white surf beat against its steep sides; then all collapsed, and the great shroud of the sea rolled on as it rolled five thousand years ago.

Herman Melville
Moby Dick, 1851

The International Whaling Commission

By the end of the Second World War entire populations of whales were at risk of extinction due to the pressures of commercial whaling. An international commission met in Washington, D. C., to consider a solution and adopted a range of measures that even today, despite extensive revisions, are still open to interpretation.

The IWC

In Washington, D. C., on 2 December 1946, the International Whaling Commission (IWC) was set up by the agreement of delegates from nineteen countries: Argentina, Australia, Brazil, Canada, Chili, Denmark, France, Great Britain, the Irish Republic, Japan, Mexico, the Netherlands, Northern Ireland, Norway, Panama, Peru, South America, the Soviet Union and the United States.

The measures adopted by the IWC aimed to achieve four basic objectives:
- to protect immature whales of all species in order to ensure future reproduction
- to set limits on the numbers and size of whales that may be taken
- to set aside protected areas, especially in the breeding grounds
- to ban the capture of any animal belonging to an endangered species

When it was first set up, the IWC was motivated by economic concerns rather than by the desire to protect endangered animals. During the first fifteen years of its existence it was a cartel established to stabilize whale oil prices.

It was only in the 1970s, when stocks fell dramatically, threatening to undermine the commercial viability of whaling as an industry, that serious discussion began regarding the protection of whales. It was not until 1986, when a ten-year moratorium on commercial whaling was made, that the IWC adopted more drastic measures to ensure the protection of whales.

The limitations of its remit

The IWC operates in much the same way as the United Nations' Security Council, and both bodies suffer from the same limitations: like the UN, the IWC can only make recommendations; it cannot impose its decisions on member countries or interfere with the sovereignty of represented governments. Article 5 of the organization's Charter stipulates that any signatory disagreeing with a protective measure or a quota recommended by the IWC will remain exempt from such a measure or quota provided that it raises its objection within ninety days. All member countries also have the right, of course, simply to leave the IWC at any time.

Since 1946 there have been numerous

'objections', and equally numerous defections. Such extreme measures are not always necessary, however: since IWC recommendations depend not on a simple majority, but on three-quarters of the votes, the whaling nations have been able to block decisions by standing together to create a minority faction.

Moreover, the representatives of the whaling nations at the IWC are more often than not the principal shareholders in companies depending for their profits on the wholescale slaughter of whales – hardly the people most suitable to entrust with their *protection*, any more than one might entrust a herd of sheep to a pack of wolves, or a lettuce field to a colony of rabbits.

The IWC has simply played a role in supervising the continuing decline in whale populations. What little progress it has made has been largely due to outside pressure from those in favour of abolishing whaling. As an organization it has served its time, and Greenpeace recommends that it should be replaced with a responsible scientific committee working exclusively on behalf of the United Nations Environment Programme.

The Cousteau Foundation has been recommending for years the creation of an internationally recognized organization, composed of experts independent of any private or state-run company, whose job would be to oversee matters relating to the world's oceans, and one of whose most urgent priorities would be precisely that of saving our whales – as a debt we owe to future generations.

The member countries

There are currently thirty-five member countries, which can be subdivided as follows:

- the whaling nations (Japan, Norway, Russia and Korea) and nations with indirect economic interests in whaling (Antigua and Barbuda, Dominica, Grenada, St Lucia, St Vincent and the Grenadines, St Kitts-Nevis, the Solomon Islands)
- countries granted a quota for subsistence purposes (Denmark for Greenland, the United States for Alaska, etc.)
- non-whaling nations or nations who have ceased their whaling activities and who support resolutions in favour of conservation and protection (Argentina, Australia, Austria, Brazil, Chili, France, Germany, Great Britain, India, Italy, Monaco, the Netherlands, New Zealand, Sweden)
- shifting nations who are unpredictable in their manner of voting (China, Finland, Ireland, Mexico, Oman, South Africa, Spain, Switzerland)

There are also a number of non-member countries which are able to observe the proceedings of the IWC but do not have the right to vote (Canada, the European Union, Iceland, Iran, Morocco, Zimbabwe), various inter-governmental organizations and a total of seventy non-governmental organizations, including organizations like Greenpeace established for the protection of the environment and others that favour whaling, including a number of associations established for the promotion of the Japanese whaling industry.

Some important dates

1982

This year was the turning point for the whaling industry. The IWC met at Brighton and decided by 25 votes to 7 (with 7 abstentions) in favour of a complete ban on commercial whaling;

the ban was due to come into force in 1986.

In addition to its nineteen founder members the IWC now had a further nineteen members including large, densely populated countries like China, India and Egypt and small island states sensitive to the need to preserve the riches of the world's oceans (the Seychelles, St Lucia, St Vincent, Antigua).

The moratorium was adopted despite pressure and counter-manoeuvres on the part of the Soviets and the Japanese, but only after a series of surreal negotiations in which the fate of the whales became entangled with sales figures for Brazilian coffee, the hazards of economic cooperation between Japan and the Seychelles, and the geopolitical and ideological repercussions of the Falklands War. For the friends of the whales the most surprising, and pleasing, twist came when Spain voted for the moratorium despite prior opposition. Japan and the Soviet Union, however, took advantage of their right to object within the prescribed ninety days, a decision in which they were supported by Norway and Peru.

1983

Brazil, Iceland, Peru and South Korea accepted the moratorium, but the Philippines – a nation who had never previously hunted whales – succumbed to pressure from Japan and announced their intention to capture two hundred Bryde's whales annually. In another unfortunate turn of events the quotas for the slaughter of bowhead whales were raised as a further concession to indigenous populations.

1984

At the IWC's annual conference, held at Buenos Aires, the Argentine government decided to raise a monument to the right whale on a small island in the Valdes peninsula – a symbolic gesture accompanying a significant reduction in authorized quotas (from 9875 between 1983 and 1984 down to 6837 between 1984 and 1985, the final year before the moratorium came into force). Brazil, Japan and the Soviet Union objected to the reduction, and Norway contrived to find a loophole in the regulations, arguing the case for a third type of hunting, neither commercial nor aboriginal but reserved for 'artisans', notably in the North Atlantic.

1985

The activities of the small handful of nations who still practised whaling (resorting to the most elaborate stratagems in defence of their actions) led to the imposition of sanctions by the United States and the countries of the European Union. On 5 April 1985 the Japanese Minister for Agriculture and Fisheries announced that his country would be disbanding its whaling fleet in 1988. Whether this was a firm commitment or simply a means of sidestepping the moratorium was unclear.

1986

When the moratorium came into force, Japan, Norway and Russia continued to object and refused to disband their fleets, and Iceland and South Korea declared that they were switching from commercial to scientific whaling. The Caribbean islands of St Lucia and St Vincent aligned themselves with the whaling nations – a dramatic shift in policy that coincided with massive local investments in aid programmes aimed at developing the islands' fishing industry. The money came from Japan.

1987

Since the proposal to introduce a moratorium on commercial whaling, Japan claimed to have converted its commercial fleet into a scientific fleet. The legal confusion surrounding the enforcement of the moratorium now made it possible for Japan, Norway and Iceland to reorganize their whaling programmes along scientific lines.

1988

Russia abandoned commercial whaling.

1989

Japan's scientific fleet hunted minke whales in Antarctica.

1990

The whaling nations sought authorization to hunt the minke whale in addition to other available species.

1991

The IWC adopted the Revised Management Procedure (RMP). Iceland ceased to be a member of the IWC. The moratorium on commercial hunting remained in force.

1992

Dominica supported the whaling nations in exchange for economic aid. Norway set up the North Atlantic Commission with the aim of weakening the authority of the IWC in the region.

1993

The island of Grenada joined the other states under Japanese protection in the IWC. Norway resumed commercial whaling. An illegal trade in whale meat was uncovered (labelled shrimp and destined for South Korea and Japan).

1994

The need for a whale sanctuary in Antarctica was recognized by the IWC but opposed by Japan. The Russian Minister for Fisheries revealed secret documents regarding the falsification of information during commercial whaling expeditions. The captain of a Norwegian whaler was prosecuted for exceeding national quotas – the first time that such a prosecution had occurred. Japan began a new scientific programme, focusing on a population of minke whales in the North Pacific.

1995

The IWC's scientific committee discovered an error in the computer programme used by the Norwegians to estimate the population of minke whales in the North Atlantic.

1997

The Japanese fleet extended its activities to Antarctica's zone VI. Analyses of DNA samples taken from meat on sale in Japan and South Korea proved that protected species were being illegally slaughtered. Japan proposed a resolution aimed at establishing a secret ballot – a strategy that would have enabled it to conceal even more closely agreements made with smaller nations.

1998

Norway raised its quota of minke whales to 670, double the number it had specified in 1993 when it had resumed commercial whaling. Japan, aided by South Korea, China and Russia, attempted to form an organization to oversee whaling in the North Pacific. Iceland announced its intention to resume whaling, and Japan sought to apply its scientific programme to a population of Bryde's whales.

Yves Cohat and Anne Collet

Substitute products

During the 19th century people made their fortunes from trading in whale products, but today a substitute exists for every product extracted from whales and substitution is often the cheaper option. As whaling becomes less and less profitable, there can be no justification for continuing the slaughter.

Collagen

This protein is extracted from the skin and bones of whales and boiled to produce gelatin, which is used in cosmetics, in food preserving and the preparation of processed meats, and in the photographic industry.
Substitutes: vegetable matter, seaweed

Sperm whale teeth

These teeth are made of ivory and are principally used in sculptures. Formerly they were also used to make buttons and dice.
Substitutes: synthetic materials, plastics

Liver and endocrine glands

Whale liver oil is richer in vitamins A and D than cod liver oil and is widely used in the pharmaceuticals industry. A number of hormones are also extracted from the endocrine glands (pituitary, pancreas and adrenal).
Substitutes: betacarotene, cod liver oil, synthetic vitamin A

Whale oil

Prior to the discovery of electricity and crude oil in the 19th century, whale oil was used for lighting purposes and in the manufacture of candles and soap. Its value has diminished today, but it continues to be used in the manufacture of explosives and materials for the chemicals and pharamaceuticals industry and in preservatives.
Substitutes: beeswax and paraffin wax, rapeseed, jojoba and flaxseed or linseed oil

Sperm oil and spermaceti

Sperm oil is refined and filtered and used in the manufacture of luxury candles and cosmetics (beauty creams, lipsticks, ointments and shaving creams). It can also be used as a lubricant for engines, machine parts and precision instruments, and in the preparation of paint solvents, printer's ink, carbon paper, plastics and even pasta.
Substitutes: linseed, castor and rapeseed oil, essential oils of lemon and orange, avocado oil, cucumber milk and jojoba oil

Bones

Whalers used to make cooking utensils, knitting needles and chess pieces from whale bones. Today the bones are reduced to powder and used in the preparation of fertilizers and animal feed.
Substitutes: cereal husks, seaweed, animal remains (in abattoirs) etc.

Skin

A gelatinous substance was formerly extracted from whale skin and used in the manufacture of bicycle seats, handbags, shoes and laces.

Substitutes: plant, (stock) animal and chemical substances

Blood

Blood is used to make fertilizers.

Substitute: fertilizer produced from vegetable matter

Tendons

Tendons are used to make strings for tennis rackets and surgical thread.

Substitute: tendons of stock animals

Meat

Whale meat is eaten in Japan – where it represents a mere 1.7 per cent of the total meat consumption – and in South Korea, Iceland, Norway, Greenland and Alaska. It is also an ingredient in dog and cat food and is fed to animals bred for their fur.

Viscera

They are cooked and dried before being turned into fertilizer. The bile is used to make dyes and varnishes.

Substitutes: fertilizers made from vegetable matter; chemical and synthetic substances

Ambergris

Ambergris ('grey amber') is a waxy substance secreted by the intestinal tract of the sperm whale. It consists mainly of stable alcohols like ambrein, a substance closely related to cholesterol. It has a sickly smell when fresh, but, as it ages, it begins to give off a strong but pleasant muskiness. Sought after as a perfume in the Middle Ages, ambergris is still used today as a fixative in the manufacture of perfumes as well as an ingredient in cosmetics and luxury soaps, and it continues to fetch a high price. 'Who would think, then,' writes Herman Melville in *Moby Dick*, 'that such fine ladies and gentlemen should regale themselves with an essence found in the inglorious bowels of a sick whale! Yet so it is. By some, ambergris is supposed to be the cause, and by others the effect, of the dyspepsia in the whale.' Ambergris also possesses antispasmodic properties and is used in the preparation of certain drugs. The largest piece of ambergris was found at the beginning of the last century in the stomach of a sperm whale: it weighed more than 450 kilos (990 pounds).

Substitute: fixative 400

TDC (Textes et Documents pour la Classe), October 1993, No. 661

Ajojoba plant.

Why whales leap

Watching a whale leap, watching that mountain of flesh – 30 tonnes and more, 15 metres (50 ft) and more – soaring out of the sea, is an incredible experience. Such behaviour, which is common among certain species, appears to have a social significance. It serves as a means of communication between whales, whether as a challenge or a signal, or simply as a game. This article remains the best researched on the subject.

A whale's leap from the water is almost certainly the most powerful single action performed by any animal. It is called breaching, a term that whalers of the 18th and 19th centuries gave this dramatic activity and that present-day investigators of the phenomenon have retained. Considering the great bulk and weight the whale must lift in breaching, one wonders why the animal does it....

The whalers of earlier centuries, searching for their quarry in slow sailing vessels, had many opportunities to observe the whales they were trying to catch. For years the anecdotes told by such men formed the basis of what was known of breaching and other kinds of whale behavior. Among the explanations of breaching they proposed, somewhat anthropomorphically, were feeding, stretching, amusement, being chased by swordfish and an 'act of defiance', which was presumably directed at the whalers.

In the past few years scientific observations of whales in the open ocean have begun to yield useful quantitative data on many aspects of their behavior, including the breach....

A leap by a humpback entails the lifting of as much biomass as would be accounted for by 485 people weighing an average of 68 kilograms (150 pounds) each. The largest humpbacks reach lengths approximating 15 meters (49 feet) and weigh 33 metric tons (72,765 pounds).

The breaches of the humpback and of other whales known to breach range from a full leap clear of the water to a leisurely surge in which only half of the body emerges....

Another activity in which certain aquatic animals intentionally jump above the surface is porpoising. The animal makes a series of horizontal leaps while traveling fast. Robert W. Blake of the University of British Columbia has calculated that by making such leaps a small whale or a dolphin minimizes frictional drag. He has also shown that large whales would not benefit in this way by porpoising, and indeed I have never seen humpbacks do it....

How much energy is a whale consuming as it makes a breach, and how much power is it developing as it leaves the surface? Using measurements

from photographs of breaching whales, I have simulated the breaching process on a small computer. In a full breach, in which most of the animal leaves the surface of the water at an angle of about 35 degrees, a 12-meter [39-foot] adult humpback breaks the surface at about 15 knots (17 miles per hour). Because that is almost the maximum speed the animal can attain, a full breach represents the extreme use of a humpback's propulsive power.

The energy necessary to make such a breach is roughly 2,500 kilocalories. The whale's resting metabolic rate is some 300,000 kilocalories per day. Hence the energy consumed in a breach is a little less than a hundredth of the animal's minimum daily calorific requirement…. One breach is therefore not a particularly significant event in the daily energy budget of a whale. A sequence of 20 breaches or more, however, consumes a good deal of energy. It is not surprising that successive breaches are weaker.

It is less easy to say why a whale breaches…. I have spent several hundred hours in small sailboats following groups of humpbacks through their daily routines. This work, together with the observations by [Roger] Payne [of the U. S. World Wildlife Fund] and others, is yielding a fairly clear picture of the circumstances in which whales breach…. The best one can do is to put forward statistically significant tendencies. What they suggest is that breaching is mainly associated with social interaction among whales, perhaps in communication and (among young whales) play.

Whales often breach when a pod containing two or more humpbacks splits into two groups or when two pods (sometimes consisting of single whales) merge. A breach also often takes place within 15 minutes of a lobtail: a thrash of the whale's flukes onto the surface of the water….

It is notable and apparently contradictory that humpbacks breach less in summer, even though groups split and merge more often then than they do in winter. Mating and calving take place in winter, however, and such social interactions are probably more important than the summer ones. Hence breaching rates are correlated not only with the number of social interactions taking place but also with their importance in the life of the whales.

An additional correlation between breaching and social activity is seen when one looks at the rates of breaching among different whale species....
The more rotund species would seem to be less likely to breach because of unfavorable hydrodynamics. It is surprising, then, that observations show that they do it more frequently.

Right whales, gray whales and humpbacks – the three best-studied rotund species – congregate in winter on traditional breeding grounds. They seldom feed there, subsisting instead on the energy stored in their thick layers of blubber. Social interactions are frequent and sometimes vigorous on those breeding grounds, and it is there that most breaching is seen.

In contrast the blue whale (*Balaenoptera musculus*), the finback (*B. physalus*) and the sei (*B. borealis*) – all slim – do not seem to frequent particular breeding grounds but remain dispersed during the winter months. This strategy probably reduces their net expenditure of energy, so that they do not need thick blubber layers. They may employ loud low-frequency sound or perhaps a monogamous social system to obtain access to mates. In any event they probably have rather few close-range social interactions.

Little is known of the social systems of the bowhead (*Balaena mysticetus*), Bryde's whale (*Balaenoptera edeni*) or the minke (*B. acutorostrata*), but the general impression among close observers is that among these baleen whales the more social species have the higher breaching rates. The sperm whale (*Physeter catodon* [now *Physeter macrocephalus*]), a toothed whale that breaches frequently, has a particularly complex social system.

What other clues emerge from investigations of the context of breaching? One unexpected finding, obtained in several independent studies, is that whales breach oftener as the wind speed rises.... Payne has speculated that whales might be using the breach as a means of communicating by sound (from the slap of reentry) when noise from wind and waves obscures their normal vocalizations.

Payne made another discovery that led him to think breaches might have a signaling function. He found that among southern right whales breaching begets breaching. In other words, the likelihood that an individual whale would breach increased when whales nearby were breaching.... Under good conditions they might be expected to hear the sound of a breach over a distance of a few kilometers. Thus the findings lend tentative support to Payne's hypothesis that breaching has a signaling function. If other whales see or hear a breach, information has been conveyed. The message would at least be that a whale has breached.

Is a breach an efficient way of conveying any other message? It makes a spectacular sight and a loud noise for observers on the surface, but most of the other whales are below the surface at the time of a breach. Even in the clearest water the limit of underwater vision is about 50 metres. Under favourable conditions, however, sound can travel quite far in seawater. The question therefore becomes whether a whale can generate louder sounds, in at least a few frequency ranges, by breaching than it can by vocalizing. Little information is available on the strength of the underwater sound produced by a breach and no information is available on whether whales try to maximize their output of sound during a breach.

A breach might also be a display intended as an act of aggression, as a challenge, as a show of strength or as a maneuver in courtship…. During many months at sea in small boats, I have never felt that any of the thousands of breaches we observed was aggressively directed toward us. Moreover, the whale can probably display aggression more effectively by administering a blow with its flukes.

A whale making a full breach is exhibiting its maximum power to any whale within sight or earshot. Hence the breach might be useful as a courtship display, a challenge or a show of strength. A female might choose a mating partner at least partially on the basis of the strength of his breach or his ability to keep up a strong output of power or sound during a sequence of breaches. Such a male would be demonstrating strength and stamina and so perhaps (indirectly) genetic fitness.

Similar correlations might make a breach useful as a challenge or a show of strength directed at other males competing for access to a particular female. Breaches by right whales and humpbacks are often seen when males are engaged in such a competition.

One must also consider the somewhat blurry concept of play. People watching an animal perform an action with no immediately obvious function tend to call it play. As a result the concept has become a catchall category for otherwise inexplicable behavior, in which breaching has often been included. Recently play has had serious attention from a number of biologists and students of animal behavior, and it is now generally regarded as a valid (but hard to define) behavioral category….

Breaching has most of the characteristics of other activities that animal behaviorists call play: it is common in social contexts, it is often done by young animals and in many instances it has no obvious function. Some investigators have speculated that a purpose of play in other young animals is to aid the development of musculature; breaching might serve this role in young whales.

The most spectacular breaches are made by the youngest whales. Right-whale, gray-whale and humpback calves begin breaching when they are only a few weeks old. The breaches are often vigorous and may run on in long sequences….

The findings I have reported and the hypotheses I have discussed do not indicate any single clear function for breaching. The evidence suggests the activity has several functions. Although there are strong correlations with sociality and breaches have characteristics that would make them effective as signals of physical prowess, no evidence conclusively supports either hypothesis.

My subjective evaluation is that breaching often serves to accentuate other visual or acoustic communication. It is a kind of physical exclamation point. Just as people raise their voice, gesticulate with their hands or jump up and down to emphasize a communication, so the whale breaches. And, like eavesdroppers, human observers usually miss the message, noting only its salient features.

Hal Whitehead
'Why Whales Leap',
Scientific American,
March 1985,
Vol. 252, No. 3

Whales do not commit suicide

Beached whales are not an uncommon sight, and researchers have been investigating the reasons for this phenomenon. Is pollution to blame, or parasitic illness, or some kind of group instinct? Rational explanations rule out suicide.

A 19th-century engraving of a beached whale.

In France discoveries of beached whales have been systematically recorded since 1972. Two or three hundred animals were found annually along the coast until the end of the 1980s, then suddenly the numbers rose to between five and nine hundred a year, dolphins found on the coasts of the Bay of Biscay accounting for the great majority of these figures. This sudden increase is unfortunately not linked to population growth, but rather to an increased mortality.

Close examination reveals that new fishing techniques, such as trawling, are responsible for these multiple deaths. Whales (true whales or whalebone whales) represent a small proportion (1 to 3 per cent) of the total numbers of cetaceans that perish in this manner, both in Europe and in the rest of the world. Whales tend to live further offshore and their populations are much smaller than those of dolphins: therefore there is a much higher probability of discovering a dead dolphin on a beach than finding a whale.

The idea that whales commit suicide is a myth that has enduring appeal – as if the ability to take its own life at will were a feature linking whales and human beings. We find it gratifying to see ourselves reflected in the creatures we love, and yet the events that have given rise to the myth of whale 'suicides' have nothing to do with (true) whales – or with suicide.

These events in fact concern several species of toothed whales (including pilot whales, orcs and sperm whales). The confusion originates from the fact that, in English, we use the term 'whale' to include all large cetaceans, both toothed whales (*Odontoceti*) and whalebone, or true, whales (*Mysticeti*). In French the term *baleine* is reserved for whalebone whales alone; but when a group of pilot whales is found beached on the coast of Australia or Scotland, the French-speaking media translates the word as *baleine pilote*, thus contributing to the confusion. The fact is that no species of true whale has ever perished in a manner that could give rise to the interpretation of suicide.

When ten, or even one hundred, live pilot whales are stranded on a beach, it is difficult to manoeuvre between this mass of large dolphins (the adults are between 4 and 6 m (13 and 20 ft) long, thrashing about in distress. A rescue team can only handle one animal at a time when trying to return the stranded animals to the sea, and will tend to select the young animals first, because they are lighter and easier to manipulate. Dolphins are thought to be fairly gregarious, however, and their group instinct is frequently stronger than the instinct of the individual. Back in the sea, the young dolphin's one aim will be to return to its family stranded onshore – an act that has come to be interpreted as suicide.

Having understood their mistake, the rescuers now know that, for their mission to succeed, they must drive several dolphins out to sea simultaneously, taking care to include a dominant animal (one of the largest) in the group.

The question remains as to why these herds of dolphins are washed ashore when they appear to be perfectly healthy? The answer is that they have probably made a navigational error due to the nature of the terrain. Indeed it has been shown that the majority of such cases occur at the head of sandy bays, which serve as a poor sounding board for the sonar on which the dolphin relies for navigation. Moreover, anomalies in the earth's magnetic field are often located in areas where dolphins have been found stranded and would appear to contribute to the animals' confusion. Mass panic is another probable factor contributing to these sad occurrences.

Anne Collet

A beached pilot whale at Roscoff in Brittany in December 1904.

SUBORDER *MYSTICETI* (BALEEN WHALES OR MYSTICETES), 13 SPECIES

FAMILY *BALAENIDAE* (RIGHT WHALES), 4 SPECIES

Balaena mysticetus (bowhead whale)
Eubalaena australis (southern right whale)
Eubalaena glacialis (North Atlantic right whale)
Eubalaena japonica (North Pacific right whale)

FAMILY *BALAENOPTERIDAE* (BALAENOPTERIDS), 7 SPECIES

Balaenoptera acutorostrata (common minke whale)
Balaenoptera bonaerensis (Antarctic minke whale)
Balaenoptera borealis (sei whale)
Balaenoptera edeni (Bryde's whale)
Balaenoptera musculus (blue whale)
Balaenoptera physalus (fin whale)
Megaptera novaeangliae (humpback whale)

FAMILY *ESCHRICHTIIDAE* (ESCHRICHTIDS), 1 SPECIES

Eschrichtius robustus (grey whale)

FAMILY *NEOBALAENIDAE* (NEOBALAENIDS), 1 SPECIES

Caperea marginata (pygmy right whale)

SUBORDER *ODONTOCETI* (TOOTHED WHALES OR ODONTOCETES), 69 SPECIES

FAMILY *DELPHINIDAE* (DELPHINIDS), 34 SPECIES

Cephalorhynchus commersonii (Commerson's dolphin)
Cephalorhynchus eutropia (Chilean dolphin)
Cephalorhynchus heavisidii (Heaviside's dolphin)
Cephalorhynchus hectori (Hector's dolphin)
Delphinus capensis (long-beaked common dolphin)
Delphinus delphis (common dolphin)
Feresa attenuata (pygmy killer whale)
Globicephala macrorhynchus (short-finned pilot whale)
Globicephala melas (long-finned pilot whale)
Grampus griseus (Risso's dolphin)
Lagenodelphis hosei (Fraser's dolphin)
Lagenorhynchus acutus (Atlantic white-sided dolphin)
Lagenorhynchus albirostris (white-beaked dolphin)
Lagenorhynchus australis (Peale's dolphin)
Lagenorhynchus cruciger (hourglass dolphin)
Lagenorhynchus obscurus (dusky dolphin)
Lagenorhynchus oliquidens (Pacific white-sided dolphin)
Lissodelphis borealis (northern right whale dolphin)
Lissodelphis peronii (southern right whale dolphin)
Orcaella brevirostris (Irrawaddy dolphin)
Orcinus orca (killer whale)
Peponocephala electra (melon-headed whale)
Pseudorca crassidens (false killer whale)
Sotalia fluviatilis (tucuxi)
Sousa chinensis (Indo-Pacific humpbacked dolphin)
Sousa teuszii (Atlantic humpbacked dolphin)
Stenella attenuata (pantropical spotted dolphin)
Stenella clymene (clymene dolphin)
Stenella coeruleoalba (striped dolphin)
Stenella frontalis (Atlantic spotted dolphin)
Stenella longirostris (spinner dolphin)
Steno bredanensis (rough-toothed dolphin)
Tursiops aduncus (Indo-Pacific bottlenose dolphin)
Tursiops truncatus (common bottlenose dolphin)

FAMILY *INIIDAE* (INIIDS), 1 SPECIES

Inia geoffrensis (boto, Amazon river dolphin)

FAMILY *KOGIIDAE* (PYGMY SPERM WHALES), 2 SPECIES

Kogia breviceps (pygmy sperm whale)
Kogia sima (dwarf sperm whale)

FAMILY *LIPOTIDAE* (LIPOTIDS), 1 SPECIES

Lipotes vexillifer (baiji, Chinese river dolphin)

FAMILY *MONODONTIDAE* (MONODONTIDS), 2 SPECIES

Delphinapterus leucas (white whale)
Monodon monoceros (narwhal)

FAMILY *PHOCOENIDAE* (PORPOISES), 6 SPECIES

Neophocaena phocaenoides (finless porpoise)
Phocoena dioptrica (spectacled porpoise)
Phocoena phocoena (harbour porpoise)
Phocoena sinus (vaquita)
Phocoena spinipinnis (Burmeister's porpoise)
Phocoenoides dalli (Dall's porpoise)

FAMILY *PHYSETERIDAE* (PHYSETERIDS), 1 SPECIES

Physeter macrocephalus (sperm whale)

FAMILY *PLATANISTIDAE* (PLATANISTIDS), 1 SPECIES

Platanista gangetica (South Asian river dolphin)

FAMILY *PONTOPORIIDAE* (PONTOPORIDS), 1 SPECIES

Pontoporia blainvillei (franciscana)

FAMILY *ZIPHIIDAE* (BEAKED WHALES), 20 SPECIES

Berardius arnuxii (Arnoux's beaked whale)
Berardius bairdii (Baird's beaked whale)
Hyperoodon ampullatus (northern bottlenose whale)
Hyperoodon planifrons (southern bottlenose whale)
Mesoplodon bahamondi (Bahamonde's beaked whale)
Mesoplodon bidens (Sowerby's beaked whale)
Mesoplodon bowdoini (Andrews' beaked whale)
Mesoplodon carlhubbsi (Hubbs' beaked whale)
Mesoplodon densirostris (Blainville's beaked whale)
Mesoplodon europaeus (Gervais' beaked whale)
Mesoplodon gingkodens (gingko-toothed beaked whale)
Mesoplodon grayi (Gray's beaked whale)
Mesoplodon hectori (Hector's beaked whale)
Mesoplodon layardii (strap-toothed whale)
Mesoplodon mirus (True's beaked whale)
Mesoplodon pacificus (Longman's beaked whale)
Mesoplodon peruvianus (pygmy beaked whale)
Mesoplodon stejnegeri (Stejneger's beaked whale)
Tasmacetus shepherdi (Shepherd's beaked whale)
Ziphius cavirostris (Cuvier's beaked whale)

This list, which includes the scientific names (in Latin) and common names (in English) of all cetaceans, was drawn up by the IWC in July 2000.

PROTECTION OF THE DIFFERENT SPECIES

ANTARCTIC MINKE WHALE

8–10 m (26–32 ft). 6–10 tonnes
World population: 600,000–1,140,000
Protection: commercial whaling banned since 1986; some 400 animals captured annually by Japan for scientific purposes, as authorized by the IWC

BLUE WHALE

24–27 m (78–88 ft), max. 36 m (118 ft). 80–160 tonnes
World population: 1500–5000
Protection: commercial whaling banned since 1967; some animals probably caught without authorization in Indonesia

BOWHEAD WHALE

13–15 m (42–49 ft), max. 19 m (62 ft). 60–110 tonnes
World population: 6000–8000
Protection: commercial whaling banned since 1935; the Inuit authorized to capture a small number of animals by the IWC

BRYDE'S WHALE

11–13 m (36–42 ft), max. 14 m (45 ft). 18 tonnes
World population: c. 90,000
Protection: commercial whaling banned since 1986; some animals probably caught without authorization in Indonesia

FIN WHALE

18–22 m (59–72 ft), max. 26 m (85 ft).
40–100 tonnes
World population: c. 120,000
Protection: commercial whaling banned since
1986; Greenland Inuit authorized to capture a
small number of animals by the IWC; some
animals probably caught without authorization
in Indonesia

GREY WHALE

10–13 m (32–42 ft). 20–35 tonnes
World population: 20,000–26,000
Protection: commercial whaling banned since
1935; the Inuit authorized to capture a small
number of animals by the IWC

HUMPBACK WHALE

13–15 m (42–49 ft), max. 17 m (55 ft).
20–50 tonnes
World population: c. 15,000
Protection: commercial whaling banned since
1966; St Vincent and the Grenadines authorized
to capture a small number of animals by the
IWC

MINKE WHALE

8–9 m (26–29 ft). 6–8 tonnes
World population: 1 million
Protection: commercial whaling banned since
1986; Greenland Inuit authorized to capture
a small number by the IWC; Japan authorized
by the IWC to capture 50–100 animals annually,
although the actual number taken probably
exceeds 500; some 300 animals captured
annually by Norway in the North Atlantic

NORTH ATLANTIC RIGHT WHALE

12–15 m (39–49 ft), max. 17 m (55 ft).
40–100 tonnes
World population: 800–1000
Protection: total since 1935

NORTH PACIFIC RIGHT WHALE

12–15 m (39–49 ft), max. 17 m (55 ft).
40–100 tonnes
World population: 200–300
Protection: total since 1935

PYGMY RIGHT WHALE

5–6 m (16–19 ft)
World population: unknown
Protection: never exploited; protected since 1986

SEI WHALE

13–15 m (42–49 ft), max. 21 m (68 ft). 20 tonnes
World population: 55,000–100,000
Protection: commercial whaling banned since
1986

SOUTHERN RIGHT WHALE

13–15 m (42–49 ft), max. 18 m (59 ft).
50–100 tonnes
World population: c. 3000
Protection: total since 1935

SPERM WHALE

10–15 m (32–49 ft), max. 18 m (59 ft).
15–45 tonnes
World population: c. 2,000,000
Protection: whaling banned since 1985

FURTHER READING

Bullen, Frank T., *The Cruise of the Cachalot*, 1898
Carwardine, Mark, *On the Trail of the Whale*, 1994
Carwardine, Mark, and Martin Camm, *Whales, Dolphins and Porpoises*, 1995
Carwardine, Mark, et al, *Whales and Dolphins: The Ultimate Guide to Marine Mammals*, 1998
Collet, A., G. Ross, and B. Saladin d'Anglure, *Baleines, un enjeu écologique*, 1999
Corrigan, Patricia, *The Whale Watcher's Guide*, 1994

Cousteau, Jacques-Yves, and Yves Paccalet, *Whales*, 1986
Darling, James, and Flip Nicklin, *With the Whales*, 1990
Darling, James, C. Nicklin, K. S. Norris, H. Whitehead and B. Wursig, *Whales, Dolphins and Porpoises*, 1995
Evans, Peter, *Whales*, 1990
—, *Dolphins*, 1994
Francis, Daniel, *A History of World Whaling*, 1990

Gaskin, D. E., *The Ecology of Whales and Dolphins*, 1982

Harrison, Richard, and Michael M. Bryden, *Whales, Dolphins, and Porpoises*, 1988

Herman, L. M. (ed.), *Cetacean Behavior: Mechanisms and Functions*, 1980

Hoyt, Eric, *The Whale Watcher's Handbook*, 1984

—, *Seasons of the Whale: Riding the Currents of the North Atlantic*, 1998

Klinowska, Margaret, *Dolphins, Porpoises and Whales of the World*, The IUCN Red Data Book, 1991

Leach, Nicky, *Discovery Travel Adventures: Whale Watching*, 1999

Leatherwood, Stephen, and Randall R. Reeves, *The Sierra Club Handbook of Whales and Dolphins*, 1983

Lien, Jon, and Steven Katona, *A Guide to the Photographic Identification of Individual Whales*, 1990

Martin, Anthony R., *Whales and Dolphins*, 1990

May, John, *The Greenpeace Book of Dolphins*, 1990

Melville, Herman, *Moby Dick*, 1851

Obee, Bruce, and Graeme Ellis, *Guardians of the Whales: The Quest to Study Whales in the Wild*, 1992

Payne, Roger, *Among Whales*, 1995

Severin, Tim, *In Search of Moby Dick*, 1999

Simmonds, Mark P., and Judith D. Hutchinson, *The Conservation of Whales and Dolphins: Science and Practice*, 1996

Thomas, J. A., R. A. Kastelein, and A. Y. Supin, *Marine Mammal Sensory Systems*, 1992

Waller, Geoffrey, *Sealife: A Complete Guide to the Marine Environment*, 1996

Whitehead, Hal, *Voyage to the Whales*, 1989

Williams, Heathcote, *Whale Nation*, 1988

INTERNET SITES

Greenpeace: http://www.greenpeace.org

International Whaling Commission: http://ourworld.compuserve.com/homepages/iwcoffice

New Bedford Whaling Museum, USA: http://www.whalingmuseum.org

Whale and Dolphin Conservation Society: www.wdcs.org

LIST OF ILLUSTRATIONS

The following abbreviations have been used:
a above; *b* below.

CHAPTER 2

CHAPTER 3

INDEX

PHOTO CREDITS

Alex Aguilar/CRMM 78. Alaska Photo/Betty Johannsen, Washington, D. C. 115. All rights reserved 14–5, 15, 16–7, 18a, 20b, 24a, 24b, 30, 37, 38–9, 44, 57, 62, 63, 74–5, 81a, 99, 106. Marc Berthier 64, 65a, 65b. Bettman Archives, New York 49a. Bios/Allan D./OSF 34a. Bios/Barnett, Arnold 22a. Bios/Chessman, Arnold 26–7. Bios/Yves Lefèvre front cover top, 30–1. Bios/C. Lotscher, P. Arnold 88. Bios/Morgan, Arnold 21. Bios/Seitre 20a, 23, 77. Bios/Still Pictures/Mark Carwardine 32, 78–9, 92. Bios/J.-P. Sylvestre 81b, 88–9. Bios/R. Valarcher 35. The Bodleian Library, University of Oxford back cover. Bryns Foto, Sandfjord, Norway 76. Roland Cat 96. J.-L. Charmet 36. Anne Collet/CRMM 12, 22b, 24–5, 28–9, 33, 34b, 75. Editions Robert Laffont 40–1, 41a, 41b. Richard Ellis, New York front cover bottom, 1, 2–3, 4–5, 6–7, 8–9. Gallimard archives 11, 14, 18b, 19, 38, 42, 52–3, 67, 97, 118. Greenpeace, Paris 82–3, 83. Greenpeace/Gleizes 93. Greenpeace/Kiryu 84–5. Greenpeace/Morgan 72, 73. Magnum/Taconis, Paris 105. Marina/Cedri, Paris 119. Muséum National d'Histoire Naturelle, Paris 13, 16. Mystic Seaport Museum, Connecticut, USA. 70–1. Norksjøfarts Museum, Oslo 42–3, 74. Old Dartmouth Historical Society, New Bedford, USA spine, 45, 46, 47, 48, 49b, 50–1, 54–5a, 54–5b, 56, 58, 60–1a, 60–1b, 62–3, 66, 68–9, 71. James Prunier 58–9. Sygma/Sandy Huffaker Jr 94–5. Sygma/Stuart Isett 90–1. Sygma/Mark Votier 80, 85, 86–7.

TEXT CREDITS

Grateful acknowledgment is made for use of material from the following work: (pp. 114–7) Adapted from 'Why Whales Leap' by Hal Whitehead. Copyright © 1985 by Scientific American, Inc. All rights reserved.

Yves Cohat (D. Litt)
is an anthropologist with specialist knowledge
of the fishing communities of northern Europe
and their strategies for survival, focusing particularly
on the ways in which social and economic
organization affects production. He is the author of
numerous articles and popular scientific books.

Anne Collet (D.Sc)
has been the director of the Centre de
Recherche sur les Mammifères Marins at La
Rochelle, on the west coast of France, since 1995.
She is a member of several national and international
scientific bodies and has taken part in numerous expeditions.
In 1998 she published *Danse avec les baleines*.

Translated from the French by Ruth Sharman

For Harry N. Abrams, Inc.
Editorial: Eve Sinaiko
Typographic designers: Elissa Ichiyasu, Tina Thompson
Cover designer: Brankica Kovrlija

Library of Congress Cataloging-in-Publication Data

Cohat, Yves, 1953–
    [Vie et mort des baleines. English]
    Whales : giants of the seas and oceans / Yves Cohat and Anne Collet.
      p. cm. — (Discoveries)
    Includes bibliographical references (p.).
    ISBN 0–8109–2982–1 (pbk.)
    1. Whales. I. Collet, Anne. II. Title. III. Discoveries (New York,
    N.Y.)

QL737.C4 C5713   2001
599.5—dc21   2001022408